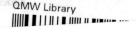

LONDON MATHEMATICAL SOCIETY LECTURE NOTE SERIES

Managing Editor: Professor N.J. Hitchin, Mathematical Institute,
University of Oxford, 24–29 St Giles, Oxford OX1 3LB, United Kingdom

The titles below are available from booksellers, or, in case of difficulty

London Mathematical Society Lecture Note Series. 287

Topics on Riemann Surfaces and Fuchsian Groups

Edited by

E. Bujalance
Universidad Nacional de Educación a Distancia, Madrid

A. F. Costa
Universidad Nacional de Educación a Distancia, Madrid

E. Martínez
Universidad Nacional de Educación a Distancia, Madrid

CAMBRIDGE
UNIVERSITY PRESS

PUBLISHED BY THE PRESS SYNDICATE OF THE UNIVERSITY OF CAMBRIDGE
The Pitt Building, Trumpington Street, Cambridge, United Kingdom

CAMBRIDGE UNIVERSITY PRESS
The Edinburgh Building, Cambridge, CB2 2RU, UK
40 West 20th Street, New York, NY 10011–4211, USA
10 Stamford Road, Oakleigh, VIC 3166, Australia
Ruiz de Alarcón 13, 28014 Madrid, Spain
Dock House, The Waterfront, Cape Town 8001, South Africa

http://www.cambridge.org

First published 2001

Printed in the United Kingdom at the University Press, Cambridge

A catalogue record for this book is available from the British Library

ISBN   0 521 00350 4 paperback

Contents

PREFACE

This book contains extended notes of most of the lectures given at the (instructional) Conference on *Topics on Riemann Surfaces and Fuchsian Groups* held in Madrid in 1998 to mark the 25th anniversary of the Universidad Nacional de Educación a Distancia. We wish to thank all contributors for their talks and texts, and in a special way to Alan F. Beardon for the excellent introduction. All papers have been refereed and we also thank to the referees for their work.

The Editors

INTRODUCTION

Riemann surfaces have played a central role in mathematics ever since their introduction by Riemann in his dissertation in 1851; for a biography of Riemann, see *Riemann, topology and physics*, Birkhäuser, 1987 by M. Monastyrsky. Following Riemann, we first consider a Riemann surface to be the natural maximal domain of some analytic function under analytic continuation, and this point of view enables one to put the theory of 'many-valued functions' on a firm foundation. However, one soon realises that Riemann surfaces are the natural spaces on which one can study complex analysis and then an alternative definition presents itself, namely that a Riemann surface is a one dimensional complex manifold. This is the point of view developed by Weyl in his classic text (*The concept of a Riemann surface*, Addison-Wesley, 1964) and this idea leads eventually on to the general theory of manifolds. These two points of view raise interesting questions. If we start with a Riemann surface as an abstract manifold, how do we know that it supports analytic functions? On the other hand, if we develop Riemann surfaces from the point of view of analytic continuation, how do we know that in this way we get all complex manifolds of one (complex) dimension? Fortunately, it turns out that these two different views of a Riemann surface are indeed identical.

We turn now to the other topic mentioned in the title of the conference, namely Fuchsian groups. One of the fundamental results in the theory of Riemann surfaces is the Uniformization Theorem. Roughly speaking, this result says that every Riemann surface is the topological quotient with respect to the action of some group Γ of analytic self-maps of one of the three classical geometries of constant curvature, namely the Riemann sphere (positive curvature), the Euclidean (or complex) plane $\mathbb{C}$ (zero curvature), and the hyperbolic plane (negative curvature). Now in all these cases the analytic self-maps are Möbius maps $z \mapsto (az + b)/(cz + d)$, and these are simple enough for the groups themselves, and their geometric action, to be analysed in depth. For topological reasons (concerning the fixed points of elements of Γ) the only group acting on the Riemann sphere that arises here is the trivial group, so the only surface that we obtain in this way is the sphere itself. For the plane $\mathbb{C}$, the group Γ is necessarily either trivial, cyclic, or a group generated by two independent translations; thus the only surfaces that we obtain in this case are the plane itself (that is, the once punctured sphere), the twice punctured sphere, and the tori. The Uniformization Theorem implies that every other Riemann surface is the quotient of the hyperbolic plane (which we may take to be the unit disc $\mathbb{D}$, or the upper half-plane $\mathbb{H}$ in the complex plane) by a group of Möbius (conformal) self-maps of the hyperbolic plane. Such groups must be discrete (otherwise the quotient structure would not be satisfactory) and these groups are the Fuchsian groups. It is immediate that because (in every case) the group Γ also acts as a group of *isometries* of the underlying geometry, the geometric structure passes down under the quotient map to the Riemann surface. We deduce that every Riemann surface comes equipped with a natural intrinsic geometry, and that in the generic case, this geometry is non-Euclidean. This, of course, is far from obvious if we consider Riemann surfaces only from the point of view of analytic continuation.

Of course, the rich theory of Riemann surfaces goes far beyond the ideas described above, and it can be seen from many other points of view, for example, group theory, combinatorics, number theory, algebraic curves, analysis, topology and so on. Let us look at some of these in a little more detail. First, the theory of algebraic curves, that is the solutions in $\mathbb{C}^2$ of some equation $P(z, w) = 0$, where P is a polynomial, is an important part of the classical theory and this is, in essence, the study of compact Riemann surfaces. Since the 1940's or so, much work has been done on Teichmüller's theory of deformation of Riemann surfaces. Suppose that we start with a given compact surface S, and that consider the class of all conformal structures on S. Each conformal structure converts the (purely topological) surface S into a Riemann surface, and Teichmüller theory is concerned with understanding how these conformal structures on S can be varied smoothly. Such arguments involve (among other things) the use of quasiconformal mappings (that is mappings whose distortion is, in a certain sense, bounded). More recently, there has been a lot of interest in hyperbolic structures on three manifolds and these can be obtained as quotients of hyperbolic 3-space by the exact analogue of a Fuchsian group acting on hyperbolic 3-space.

The notes in this text provide a significant cross-section of different aspects of Riemann surfaces, and they are mostly expository papers complete with references. As most of these contributions deal largely (and sometimes only) with compact surfaces, a brief discussion of compact surfaces might be found helpful here. Suppose that a Riemann surface R is the quotient of the hyperbolic plane by a Fuchsian group Γ. Then we can construct a fundamental region (or domain) $\mathcal{D}$ for Γ, and the Γ-images of $\mathcal{D}$ tesselate the hyperbolic plane. Now (providing that the construction of $\mathcal{D}$ is 'sensible'; for example, a convex hyperbolic polygon with side-pairings), $\mathcal{D}$ will be compact if and only if R is compact. Sometimes one is interested in those surfaces R that are a compact surface from which a finite number of punctures have been removed (this corresponds to $\mathcal{D}$ having finite hyperbolic area). More general still are the surfaces of finite type that are obtained by removing a finite number of closed discs (including punctures) from a compact surface, and these are the surfaces for which $\mathcal{D}$ is a finite sided hyperbolic polygon or, equivalently, for which Γ is finitely generated.

The first Chapter of this text is concerned only with the geometry of a Riemann surface. As a Riemann surface is the quotient of the hyperbolic plane by a Fuchsian group G (of hyperbolic isometries), the hyperbolic metric projects to a metric of constant negative curvature on the Riemann surface, and we call this the hyperbolic metric of the surface. We can now study the geometry of the Riemann surface in terms of this metric by studying the geometry of the action of G in the hyperbolic plane, and then projecting the results down to the surface. A Fuchsian group is discrete (roughly speaking, its elements are not too close together), and there are universal bounds on this separation in terms of the geometric action. These bounds give, for example, universal bounds on the lengths of certain closed loops on a Riemann surface.

Chapter 2 is concerned with arithmetic Fuchsian groups, and the way in which the number-theoretic data that defines an arithmetic group can be used to yield group-theoretic and geometric information. The groups considered here have a fundamental domain of finite hyperbolic area (such groups are known as lattices), the most familiar example being the Modular group $SL(2, \mathbb{Z})$. Starting with the modular group as the subring of matrices with integral entries in the algebra of matrices with entries in $\mathbb{Q}$, the construction is generalized by replacing $\mathbb{Q}$ by a finite extension of $\mathbb{Q}$. Quaternion algebras are then introduced, and a number of deep results on arithmetic Fuchsian groups are discussed.

The notion of a Belyi function, and Belyi's Theorem, are introduced and studied in Chapter 3. A function f is meromorphic on a compact Riemann surface R if it is an analytic map from R to the Riemann sphere. Any such function has a multiplicity, or degree, n, say, (in the sense that for any w on the sphere there are exactly n solutions of $f(z) = w$, $z \in R$, where we count multiple solutions in the usual way). For certain values of w the cardinality of the set $f^{-1}(\{w\})$ may be strictly less than n; such w are the critical values of the map f. A critical point of f is a point z in R at which f is not locally injective, and the set of critical values of f is just the set of images of the critical points of f. A meromorphic function f is a Belyi function if all of its critical values lie in the set $\{0, 1, \infty\}$. It is not obvious that such functions should have interesting properties but they do, and Belyi proved that a compact Riemann surface R supports a Belyi function if and only if it is isomorphic to the Riemann surface of some curve $P(z, w) = 0$ whose coefficients are algebraic numbers. This chapter contains a discussion of some of the consequences of these important notions of Belyi.

The set of meromorphic functions on a Riemann surface R is a field $\mathcal{M}(R)$, and Chapter 4 contains a discussion of the connection between compact Riemann surfaces and algebraic function fields of one variable over $\mathbb{C}$. This connection is studied by considering discrete valuation rings, and in particular, the valuation ring obtained from a meromorphic function on R by its (local) degree or multiplicity at a point of R. This chapter also contains a discussion of the celebrated Riemann-Roch Theorem, and illustrations of computational techniques that may be used to determine the genus, the holomorphic differentials, and the automorphism group of a particular surface.

An anti-analytic map is a map which preserves angles but reverses orientation (for example, an analytic function of $\bar{z}$), and such maps have many properties in common with analytic maps. A reflection of the hyperbolic plane across a hyperbolic geodesic is a map of the form $z \mapsto (a\bar{z}+b)/(c\bar{z}+d)$, and one can generalize the notion of a Fuchsian group to include such mappings. Such groups are called non-Euclidean crystallographic groups (NEC groups). A Klein surface is a Riemann surface together with an anti-analytic involution of itself, and compact Klein surfaces are studied in Chapters 5 and 6. A general discussion of anti-analytic involutions of a compact Riemann surface, including a discussion of NEC groups, is given in Chapter 5, and in Chapter 6 the classification of compact Klein surfaces up to a natural isomorphism is considered.

In Chapter 7, the moduli spaces of real algebraic curves are studied. Here the reader can find (among other things) a discussion of the topology and geometry of real algebraic curves along with a discussion of Beltrami differentials, quasiconformal mappings and Teichmüller spaces. The Teichmüller space of a compact Riemann surface of genus g is a complex manifold of complex dimension $3g - 3$, and this result, as well as the group of holomorphic automorphisms of the Teichmüller space, are considered here.

Chapter 8 is concerned only with compact Riemann surfaces, say of genus g. The vector space of holomorphic differential 1-forms on the surface is of dimension g and, by considering the periods of a basis of 1-forms over certain 1-cycles on R, one obtains a period matrix for the surface. The way in which the period matrix changes when one changes the bases of 1-forms, and 1-cycles, is discussed, and this leads to the description of the Siegel upper half-space, namely the space of $g \times g$ symmetric complex matrices Z with Im$[Z]$ positive definite. The change of basis of the 1-cycles leads to mappings of the form $(aZ + b)(cZ + d)^{-1}$, where the a, b, c and d are also matrices. Various properties of the period matrices are then given, and the surfaces of genus two, three and four are discussed in greater detail.

Finally, in Chapter 9, Hurwitz spaces are discussed. Each compact Riemann surface R, say of genus g, and each positive integer n, give rise to the space $H_{g,n}$ of all functions f that are meromorphic and of degree n on R. The branch points (or critical points) give local coordinates on this space, and the Hurwitz spaces are the spaces $H_{g,n}$ and their natural (and interesting) subspaces. These spaces have important physical applications, and this paper contains a discussion of some of their topological properties.

A. F. Beardon

THE GEOMETRY OF RIEMANN SURFACES

A. F. BEARDON

1. Riemann Surfaces

A *surface* S is a Hausdorff connected topological space with a collection of maps (φ_j, N_j) such that

(1) the sets N_j form an open cover of S, and

(2) φ is a homeomorphism of N_j onto $\varphi_j(N_j)$, where $\varphi_j(N_j)$ is an open subset of the complex plane $\mathbb{C}$.

A surface $\mathcal{R}$ is a *Riemann surface* if, in addition,

$$\varphi_i \varphi_j^{-1} : \varphi_j(N_i \cap N_j) \to \mathbb{C}$$

is a complex analytic function whenever $N_i \cap N_j \neq \emptyset$.

One of the easiest ways to construct a Riemann surface is as the quotient with respect to a suitable group action. For example, if G is the group generated by $z \mapsto z + 1$, then $\mathbb{C}/G$ is the punctured plane $\mathbb{C}\backslash\{0\}$, and $\mathbb{H}/G$ (where $\mathbb{H}$ is the upper half-plane) is the punctured disc $\{z : 0 < |z| < 1\}$, and in each case the quotient map is $z \mapsto \exp(2\pi i z)$. These two examples give Riemann surfaces which are the quotient of the Euclidean plane $\mathbb{C}$, and the hyperbolic plane $\mathbb{H}$, respectively (we shall discuss the hyperbolic plane in detail later). More generally, if G is any discrete group acting on $\mathbb{C}$ or $\mathbb{H}$ then the quotient by G is a Riemann surface. The deep, and powerful, Uniformisation Theorem says that (apart from the sphere) the converse is true.

The Uniformisation Theorem. *If a Riemann surface is homeomorphic to a sphere then it is conformally equivalent to the Riemann sphere. Any Riemann surface $\mathcal{R}$ that is not homeomorphic to a sphere is conformally equivalent to a quotient of the form $\mathbb{C}/G$, or $\mathbb{H}/G$, where G is a discrete group of conformal isometries acting without fixed points on $\mathbb{C}$, or on $\mathbb{H}$. Further, G is isomorphic to the fundamental group of $\mathcal{R}$.*

An isometry of $\mathbb{H}$ is a Möbius map, and it said to be *elliptic* if it has a fixed point in $\mathbb{H}$. The assumption that a group G is free of elliptic elements is a very severe restriction on the group; for example, we shall see later that in almost every case, if G is a group of isometries of $\mathbb{H}$, and if G has no elliptic elements, then G is discrete and $\mathbb{H}/G$ is a Riemann surface.

The proof and ideas surrounding the Uniformisation Theorem give much more information than this. First, one can easily classify the groups G acting on $\mathbb{C}$ that arise here, and the only Riemann surfaces of the form $\mathbb{C}/G$ are the plane $\mathbb{C}$, the punctured plane $\mathbb{C}\backslash\{0\}$, and the tori. Thus, *essentially every Riemann surface is of the form* $\mathbb{H}/G$ *for some group G of isometries of the hyperbolic plane* $\mathbb{H}$; such surfaces are known as *hyperbolic Riemann surfaces*. These facts imply that each hyperbolic Riemann surface $\mathcal{R}$ inherits, by projection from $\mathbb{H}$, its own hyperbolic geometry, and it is this geometry that we wish to study here. We shall show, for example, that in this natural metric on $\mathcal{R}$,

(1)   if γ_1 and γ_2 are loops of hyperbolic lengths ℓ_1 and ℓ_2, respectively, going round a handle on $\mathcal{R}$ as illustrated in Figure 1.1, then

$$\sinh \tfrac{1}{2}\ell_1 \sinh \tfrac{1}{2}\ell_2 \geq 1,$$

and

(2)   if σ is a loop of hyperbolic length ℓ on $\mathcal{R}$ that crosses itself (see Figure 1.1), then

$$\sinh \tfrac{1}{2}\ell \geq 1.$$

Roughly speaking, (1) shows that given any 'handle' on any hyperbolic Riemann surface, if the 'handle' is 'thin' it must be 'long', and if it is 'short' then it must be 'thick'. Notice that these results give *uniform constraints on the class of all hyperbolic Riemann surfaces*; that is, the bounds given here *do not depend on $\mathcal{R}$*.

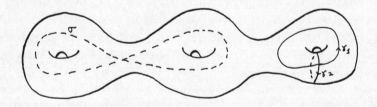

Figure 1.1

In Sections 2, 3 and 4 we study the hyperbolic plane $\mathbb{H}$ (and an equivalent model, the unit disc $\mathbb{D}$), and its group of isometries. In Section 5 we study elementary groups; as their name suggests, these groups are of little interest. Next, in Sections 6-11 we study groups of isometries that have no elliptic elements. Finally, in Section 12, we derive some universal geometric facts about Fuchsian groups and hyperbolic Riemann surfaces. Throughout, the arguments are entirely geometric, and 2×2 matrices play only a minimal part in our discussion.

2. The hyperbolic plane

Let $\mathbb{D}$ be the unit disc, and $\mathbb{H}$ be the upper half-plane; thus

$$\mathbb{D} = \{z \in \mathbb{C} : |z| < 1\}, \quad \mathbb{H} = \{x + iy : y > 0\}. \tag{2.1}$$

The Möbius maps that leave $\mathbb{D}$ invariant are of the form

$$z \mapsto \frac{az + \bar{c}}{cz + \bar{a}}, \quad |a|^2 - |c|^2 = 1, \tag{2.2a}$$

while the maps that leave $\mathbb{H}$ invariant are of the form

$$z \mapsto \frac{az + b}{cz + d}, \quad a, b, c, d \in \mathbb{R}, \ ad - bc = 1. \tag{2.2b}$$

The sets $\mathbb{D}$ and $\mathbb{H}$ are the two most common models for the hyperbolic plane. From a geometric point of view, $\mathbb{D}$ is usually the best model to use; however, the algebra is often easier in the model $\mathbb{H}$, and the reader must be familiar with both models. The point ∞ appears to play a special role in the model $\mathbb{H}$ (although of course it does not), and most of our arguments can be applied to either $\mathbb{D}$ or to $\mathbb{H}$. *We shall use $\mathcal{H}$ to denote either of these models.*

Each of the models (2.1) is equipped with a Riemannian metric of constant negative curvature, namely

$$ds = \frac{2|dz|}{1 - |z|^2}, \quad ds = \frac{|dz|}{y} \tag{2.3}$$

for $\mathbb{D}$ and $\mathbb{H}$, respectively. We shall not use the concept of curvature although in fact it is the negative curvature of the metric that gives hyperbolic geometry its own particular character. Given the metric ds in (2.3), we define the hyperbolic length $\ell(\gamma)$ of a curve γ to be

$$\ell(\gamma) = \int_\gamma ds,$$

and then the *hyperbolic distance*, say $\rho(z, w)$, between two points z and w is defined to be the infimum of $\ell(\gamma)$ taken over all curves γ in $\mathcal{H}$ that join z to w.

Any Möbius map of $\mathbb{D}$ onto itself, or of $\mathbb{H}$ onto itself, is an isometry with respect to the appropriate metric in (2.3); this is a direct consequence of the formulae

$$|g'(z)| = \frac{1 - |g(z)|^2}{1 - |z|^2}, \quad |g'(z)| = \frac{\text{Im}[g(z)]}{\text{Im}[z]}$$

which follow directly from (2.2a) and (2.2b) when g is a self-map of $\mathbb{D}$ and $\mathbb{H}$, respectively. It is not hard to show that an orientation-preserving isometry of $\mathcal{H}$ is a Möbius map of $\mathcal{H}$ onto itself, and even more is true, namely that any Möbius map between $\mathbb{D}$ and $\mathbb{H}$ (in either direction) is an isometry with respect to the appropriate hyperbolic metrics. This fact means that we can transfer ideas and formulae between $\mathbb{D}$ and $\mathbb{H}$ by means of any Möbius map that maps one onto the other; the Möbius map $z \mapsto (z - i)/(z + i)$, for example, maps $\mathbb{H}$ onto $\mathbb{D}$ but there are many other maps that could be used here.

Because any Möbius map between $\mathbb{D}$ and $\mathbb{H}$ is an isometry, we can use the same symbol, namely $\rho(z, w)$, for the hyperbolic distance between the points z and w in $\mathcal{H}$ (which can be $\mathbb{D}$, or $\mathbb{H}$). This distance is known explicitly, and is given by

$$\rho(z, w) = \log\left(\frac{1 + \tau(z, w)}{1 - \tau(z, w)}\right), \tag{2.4}$$

where

$$\tau(z, w) = \left|\frac{z - w}{1 - z\bar{w}}\right|, \quad \tau(z, w) = \left|\frac{z - w}{z - \bar{w}}\right|,$$

in $\mathbb{D}$ and $\mathbb{H}$, respectively. Note that if we use (2.4) to write $\tau(z, w)$ in terms of $\rho(z, w)$ we obtain (in both models) the equivalent statement that

$$\tau(z, w) = \tanh \tfrac{1}{2}\rho(z, w).$$

Although we have the explicit formula (2.4) for the hyperbolic distance between two points, it is rarely used in this form. The reason is that the trigonometric formulae in hyperbolic geometry always involve the hyperbolic functions $\sinh$ and $\cosh$, and so we must expect these functions to arise naturally in any formula that is in its simplest form. For example, Pythagoras' theorem in hyperbolic geometry is

$$\cosh a \cosh b = \cosh c,$$

where the given right-angled (hyperbolic) triangle has hypothenuse of length c and the other sides of lengths a and b. If the angle opposite the side of length a is α then, corresponding to the Euclidean formula $\sin \alpha = a/c$, we now have

$$\sin \alpha = \frac{\sinh a}{\sinh c}.$$

With these comments in mind, we give a more useful form of (2.4), namely

$$\sinh \tfrac{1}{2}\rho(z,w) = \frac{|z-w|}{(1-|z|^2)^{1/2}(1-|w|^2)^{1/2}}, \qquad (2.5)$$

for $\mathbb{D}$, and

$$\sinh \tfrac{1}{2}\rho(z,w) = \frac{|z-w|}{2[\mathrm{Im}(z)\mathrm{Im}(w)]^{1/2}} \qquad (2.6)$$

for $\mathbb{H}$. Each of these formulae is equivalent to, but much more useful than, (2.4).

We end this section with a brief description of the hyperbolic gemetry of $\mathcal{H}$. The geodesics in $\mathcal{H}$ are the arcs of Euclidean circles (and straight lines) that are orthogonal to the boundary $\partial \mathcal{H}$ of $\mathcal{H}$. The *hyperbolic circles* (the circles defined with respect to the hyperbolic metric ρ) are also Euclidean circles, and this means that the topology induced by the metric ρ on $\mathcal{H}$ is the Euclidean topology. If C is a circle in $\mathcal{H}$, then the Euclidean and hyperbolic centres of C coincide when, and only when, $\mathcal{H} = \mathbb{D}$ and this centre is the origin; the Euclidean and hyperbolic radii of C are always different.

The metrics (2.3) have constant curvature -1 and although we shall not use this fact (nor define what it means), the reader should be aware of the most significant consequence of it, namely that in the hyperbolic plane two (different) geodesics separate from each other very rapidly. To illustrate this in a concrete situation, observe that Pythagoras' Theorem shows that in a 'large' right-angled triangle we have (approximately) $a + b = c + \log 2$ (so that from a metric point of view the triangle appears to be very 'flat'). The explanation of this is that the two geodesic rays which form the sides leaving the right angle diverge from each other so rapidly that eventually they seem to be moving in directly opposite directions; hence the sum of the two lengths is approximately the length of the third side.

Finally, this rapid separation of geodesics appears in another form: the hyperbolic length of a hyperbolic circle of hyperbolic radius r is $2\pi \sinh r$ (which grows exponentially with r) as compared with $2\pi r$ (of linear growth) in the Euclidean case. In every sense, the hyperbolic plane has much more 'space' near its 'boundary at ∞' than does the Euclidean plane; in particular, this accounts for the existence of vary many more discrete isometry groups of $\mathcal{H}$ than there are of $\mathbb{C}$.

3. Isometries of the hyperbolic plane

We begin by discussing the conformal isometries of $\mathcal{H}$ from a geometric point of view, but first we give a warning about the terminology. There are three classes of conformal isometries of the hyperbolic plane $\mathcal{H}$, namely the hyperbolic isometries, the parabolic isometries and the elliptic isometries (which are defined in terms of fixed points). Because of this terminology we must avoid referring to an isometry of the hyperbolic plane $(\mathcal{H}, \rho)$ as simply a 'hyperbolic isometry' since there would then be an ambiguity about the meaning of this phrase. We shall now describe the geometric action of the three types of isometries, and we prefer to emphasize their *common features* rather than their differences.

(1)   *The hyperbolic isometries of $\mathcal{H}$*

A hyperbolic isometry f has two (distinct) fixed points on $\partial \mathcal{H}$. Each circular arc joining the two fixed points is invariant under f, and is called a *hypercycle* of f. Exactly one hypercycle is a hyperbolic geodesic, and this is called the *axis* A_f of f. We should think of f as being a translation of the hyperbolic plane along A_f. The simplest algebraic form of these isometries is when $\mathcal{H} = \mathbb{H}$ and the fixed points are 0 and ∞; then $f(z) = kz$, where $0 < k < 1$ or $1 < k$. The most general hyperbolic isometry is conjugate (by a Möbius map) to this map for some suitable k; more precisely, if g is a hyperbolic isometry of $\mathbb{D}$, say, then there is some Möbius map φ of $\mathbb{D}$ onto $\mathbb{H}$, and some suitable value of k, such that $g = \varphi^{-1} \circ f \circ \varphi$.

(2)   *The elliptic isometries of $\mathcal{H}$*

An elliptic isometry f has two fixed points α and β, say, with α in $\mathcal{H}$, β in the exterior of $\mathcal{H}$, and α and β inverse points with respect to $\partial \mathcal{H}$ (if $\mathcal{H} = \mathbb{H}$ then $\beta = \overline{\alpha}$; if $\mathcal{H} = \mathbb{D}$ then $\beta = 1/\overline{\alpha}$). Each (Euclidean) circle that has α and β as inverse points is invariant under f, and is called an *oricycle* of f. In fact, the oricycles in $\mathcal{H}$ are the hyperbolic circles with centre α, and f acts as a rotation of $\mathcal{H}$ about the point α. The simplest algebraic form of these isometries is when $\mathcal{H} = \mathbb{D}$, $\alpha = 0$ and $\beta = \infty$; then $f(z) = kz$, where $|k| = 1$ (and $k \neq 1$). The most general elliptic isometry is conjugate (by a Möbius map) to this map for some suitable k.

(3)   *The parabolic isometries of $\mathcal{H}$*

A parabolic isometry f has exactly one fixed point ζ, say, in the Riemann sphere, and this lies on $\partial \mathcal{H}$. We can think of f as the limiting case of either of the two previous types when their two fixed points have moved together, and for this reason parabolic isometries were known classically as 'limit rotations'. Each circle that is internally tangent to $\partial \mathcal{H}$ at ζ is invariant under f, and is called a *horocycle* for f. The simplest algebraic form of these isometries is when $\mathcal{H} = \mathbb{H}$ and $\zeta = \infty$; then $f(z) = z + a$ for some a in $\mathbb{R}$. The most general parabolic isometry is conjugate (by a Möbius map) to this map for some suitable value of a.

We now recall the forms (2.2a) and (2.2b) of the isometries of the two models $\mathbb{D}$ and $\mathbb{H}$

of the hyperbolic plane. It is well known that any Möbius map can be represented (up to a factor ± 1) by a 2×2 matrix of determinant 1, and that any such matrix determines a Möbius map. Thus, from an algebraic point of view, we can consider the isometries of the hyperbolic plane to be 2×2 matrices. In particular, the group of conformal isometries of the model $\mathbb{H}$ is $\mathrm{SL}(2, \mathbb{R})/\{\pm I\}$, where, as usual, $\mathrm{SL}(2, \mathbb{R})$ denotes the group of 2×2 real matrices with determinant 1, and I denotes the corresponding identity matrix.

Note that whereas the trace of a matrix is well-defined, the trace of a Möbius map is only defined to within a factor ± 1, and for this reason we should normally only talk of the *square of the trace* of a Möbius map. One important exception to this rule is that for given isometries f and g, the trace of a commutator $[f, g]$ ($= fgf^{-1}g^{-1}$) is independent of the choice of the matrices for f and g. It is well known that if g is an isometry of $\mathcal{H}$ (so that g is a Möbius map), then

(a) g is hyperbolic if and only if $\mathrm{trace}^2(g) > 4$;

(b) g is parabolic, or the identity I, if and only if $\mathrm{trace}^2(g) = 4$;

(c) g is elliptic if and only if $0 \leq \mathrm{trace}^2(g) < 4$.

Of course, in all cases, g is conjugate to a Möbius map with a real matrix so that $\mathrm{trace}^2(g)$ is real and hence non-negative. Although (a), (b) and (c) provide a simple arithmetic test to determine the type of a given isometry, this test is not very useful in geometrical discussions and we shall now discuss how to classify isometries by a geometric method that is based on the idea of a reflection across a hyperbolic geodesic.

First, we must describe what we mean by reflection across a Euclidean circle. Given a Euclidean circle C (or a straight line) we have a notion of inverse points with respect to this circle. We prefer the geometric definition of inverse points, so we say that the two points z and z' (not on C) are *inverse points* with respect to C if and only if every Euclidean circle through z and z' is orthogonal to C. The map $z \mapsto z'$ (taking z to its inverse point z') is an involution (that is, $(z')' = z$), and it extends naturally to C so as to fix every point on C. This map is the *reflection* (or the *inversion*) across C. Notice that as Möbius maps map circles to circles and preserve orthogonality, it is trivially true that if z and z' are inverse points with respect to C, then, for any Möbius map g, the points $g(z)$ and $g(z')$ are inverse points with respect to $g(C)$. Notice also that if C is the circle given by $|z| = r$ then the reflection across C is the map

$$z \mapsto r^2/\bar{z}. \tag{3.1}$$

The *reflection across a hyperbolic geodesic* γ is, by definition, the reflection across the unique Euclidean circle C that contains γ. As examples, if $\mathcal{H} = \mathbb{D}$ and γ is the real diameter $(-1, 1)$, then the reflection across γ is $z \mapsto \bar{z}$. If $\mathcal{H} = \mathbb{H}$ and γ is the geodesic that lies on the imaginary axis, then the reflection across γ is given by $z \mapsto -\bar{z}$. It is clear from this description that the reflection across a geodesic is an (anti-conformal) isometry of $\mathcal{H}$; in particular, the composition of two reflections is a (conformal) isometry of $\mathcal{H}$.

Finally, if g is the reflection across the geodesic γ, we say that γ is the *axis* of g; thus the axis of g is simply the set of fixed points of g. Also, given a geodesic γ in $\mathcal{H}$, we use R_γ to denote the reflection across γ. We can now describe the classification of isometries in terms of the composition of two reflections.

(1) *The hyperbolic isometries of $\mathcal{H}$*

We begin with an example. Consider the two geodesics α and β in $\mathbb{H}$ given by the circles $|z| = r_\alpha$ and $|z| = r_\beta$, respectively. Then R_α and R_β can be found from (3.1) and we find that

$$R_\beta \circ R_\alpha(z) = \left(\frac{r_\beta}{r_\alpha} \right)^2 z$$

which is of the form $z \mapsto kz$. This is a hyperbolic isometry with axis along the imaginary axis (which, significantly, is the only Euclidean circle that is orthogonal to α and β). Moreover, if z lies on this axis then

$$\rho\big(R_\beta R_\alpha(z), z\big) = 2\log(r_\beta/r_\alpha) = 2\mathrm{dist}(\alpha, \beta), \tag{3.2}$$

where $\mathrm{dist}(\alpha, \beta)$ is the (shortest) hyperbolic distance between the geodesics α and β.

As any hyperbolic isometry is conjugate to an isometry of the form $z \mapsto kz$ acting on $\mathbb{H}$, this example is, in fact, quite general. Thus if α and β are now any geodesics in $\mathcal{H}$ whose closures in the extended complex plane are disjoint, then reflection in one followed by a reflection in the other is a hyperbolic isometry whose axis is along the unique Euclidean circle that is orthogonal to both α and β. Further, as the two outer terms in (3.2) are invariant under isometries of $\mathcal{H}$, (3.2) remains valid in this general case.

Finally, the converse of this argument is true. Given any hyperbolic isometry g with axis γ, say, let $d = \rho(g(z), z)$, where $z \in \gamma$ (this is independent of the choice of z on γ; see Section 4). Now let α and β be *any* two geodesics which are orthogonal to γ and which are at a distance $\frac{1}{2}d$ apart (measured along γ). Then g is $R_\alpha \circ R_\beta$, or $R_\beta \circ R_\alpha$, depending on the labelling of α and β. In any event, the unordered pair $\{g, g^{-1}\}$ is the same as the unordered pair $\{R_\alpha \circ R_\beta, R_\beta \circ R_\alpha\}$.

(2) *The elliptic isometries of $\mathcal{H}$*

First, let α and β be two geodesics in $\mathbb{D}$ that cross at the origin. Then the hyperbolic reflection in each of these coincides with the Euclidean reflection, so that the composition $R_\alpha R_\beta$ is a Euclidean rotation about the origin, and hence also a hyperbolic rotation about the origin. As every elliptic isometry is conjugate to some map of the form $R_\alpha R_\beta$, we see that the composition of two reflections in any pair of crossing geodesics (in $\mathcal{H}$) is an elliptic isometry, and that every elliptic isometry can be expressed in this form.

(3) *The parabolic isometries of* $\mathcal{H}$
Briefly, this case is similar to the elliptic case except that here, the two (different) geodesics α and β have a common endpoint on the boundary of $\mathcal{H}$. A special case of this is where $\mathcal{H} = \mathbb{H}$, and the two geodesics end at ∞ (and so are vertical lines in $\mathbb{R}^2$). In this case the composition of the reflections is a translation, say $z \mapsto z + a$, where $a \in \mathbb{R}$. Further, any parabolic isometry is conjugate to this (for a suitable a), so the description just given is a description of the most general parabolic isometry.

4. The displacement function of an isometry
The *displacement function* of a conformal isometry f of $\mathcal{H}$, namely the function

$$z \mapsto \sinh \tfrac{1}{2}\rho\big((z, f(z))\big). \tag{4.1}$$

is extremely important, and we shall now discuss this in detail for each of the three types of isometries.

(1) *The hyperbolic isometries of* $\mathcal{H}$
Let f be a hyperbolic isometry. We shall show that there is a positive number T_f such that if z is any point on the axis $\mathcal{A}_f$ then $\rho\big(z, f(z)\big) = T_f$; we express this by saying that z is moved a distance T_f by f, and the number T_f is called the *translation length* of f. The distance moved by a general point z of $\mathcal{H}$ is a simple function of its distance $\rho(z, \mathcal{A}_f)$ from the axis $\mathcal{A}_f$ and of the translation length T_f, and is given by the beautiful formula

$$\sinh \tfrac{1}{2}\rho\big(z, f(z)\big) = \sinh(\tfrac{1}{2}T_f) \cosh \rho(z, \mathcal{A}_f). \tag{4.2}$$

Of course, this shows that if $z \in \mathcal{A}_f$ then it is moved a distance T_f by f, and, more generally, *that every point z is moved a distance of at least T_f.*

The proof of (4.2) is surprisingly easy. As every term in the formula is invariant under a Möbius map, we may assume that the hyperbolic isometry acts on $\mathbb{H}$ and that $f(z) = kz$, where $k > 1$. If y is real, then $f(iy) = iky$ so that

$$\rho\big(iy, f(iy)\big) = \rho(iy, iky) = \log k,$$

which is independent of y, so that we must have $T_f = \log k$. Also, as $\text{trace}^2(f) = (k^{1/2} + k^{-1/2})^2$, we have

$$|\text{trace}(f)| = 2 \cosh \tfrac{1}{2} T_f.$$

The proof of (4.2) for a general z is now a simple application of (2.6) with w replaced by kz.

(2)　*The elliptic isometries of $\mathcal{H}$*

Let f be an elliptic isometry. In this case the formula for the displacement function involves the angle of rotation, say θ, of f and the distance of z from the fixed point ζ of f. As we will not have any use for this case, we merely state the result, namely that

$$\sinh \tfrac{1}{2}\rho\big(z, f(z)\big) = |\sin(\theta/2)| \sinh \rho(z, \zeta). \tag{4.3}$$

(3)　*The parabolic isometries of $\mathcal{H}$*

A discussion of the displacement function of a parabolic isometry is a little more subtle than the other two cases because there is no conjugation invariant parameter (corresponding to the translation length of a hyperbolic isometry, or the angle of rotation of an elliptic isometry) that is attached to a parabolic transformation. Indeed, the two transformations $z \mapsto z + a$ and $z \mapsto z + b$, where a and b are real, are conjugate in the conformal isometry group of $\mathbb{H}$ providing only that a and b have the same sign.

Let γ be any geodesic ending at the fixed point ζ of f (recall that $\zeta \in \partial\mathcal{H}$), and let η be the other endpoint of γ. Then $f(\gamma)$ is the geodesic with endpoints ζ and $f(\eta)$, where $f(\eta) \neq \eta$, and there is a geodesic, say γ', joining η to $f(\eta)$. Note that the three geodesics γ, $f(\gamma)$ and γ' form the sides of an ideal triangle (that is, a triangle which has all of its vertices on $\partial\mathcal{H}$). Next, there is a unique horocycle, say H_f, based at ζ that is tangent to γ' and we call this the *canonical horocycle* of f (see Figure 4.1). The horocycle H_f divides $\mathcal{H}$ into two parts; the *inside* of H_f is the part whose boundary meets $\partial\mathcal{H}$ only at ζ, and the *outside* of H_f is the part whose boundary contains $\partial\mathcal{H}$.

It is important to note that the canonical horocycle defined above does *not* depend on the initial choice of the geodesic γ. To see this, we carry out the construction in the case when $\mathcal{H} = \mathbb{H}$ and $f(z) = z + 1$; then, the canonical horocycle is given by the Euclidean line $y = 1/2$ regardless of the choice of γ. Given that $f(z) = z + 1$, and that the canonical horocycle is $y = 1/2$, it is immediate from (2.6) that if $z \in H_f$, then

$$\sinh \tfrac{1}{2}\rho\big((z, f(z)\big) = 1.$$

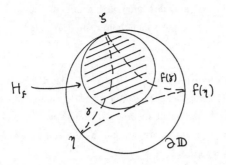

Figure 4.1

In fact, this equation (in z) defines the canonical horocycle and, as it remains invariant under conjugation, it holds for every parabolic isometry of $\mathcal{H}$.

If we still assume that $f(z) = z + 1$ and take any z in $\mathbb{H}$ with $z = x + iy$, then (2.6) shows that

$$\sinh \tfrac{1}{2}\rho\big(z, f(z)\big) = 1/2y,$$

and as

$$\rho(z, H_f) = \big|\log 2y\big| = \begin{cases} \log(2y) & \text{if } y > 1/2, \\ \log(1/2y) & \text{if } y < 1/2, \end{cases} .$$

we have now proved the following result.

Lemma 4.1. *Let f be a parabolic isometry with canonical horocycle H_f, and let z be any point in $\mathcal{H}$. Then*

$$\sinh \tfrac{1}{2}\rho\big(z, f(z)\big) = \exp\big[\varepsilon(z)\rho(z, H_f)\big],$$

where $\varepsilon(z)$ is 1 if z is on or outside H_f, and -1 if z is inside H_f.

Notice that if z is inside H_f but at a large distance from it, then $\rho\big(z, f(z)\big)$ is very small. Thus, in contrast to hyperbolic isometries, for a parabolic isometry f we have $\inf_z \rho\big(z, f(z)\big) = 0$. In fact, if we write

$$\mu_f = \inf\{\rho\big(z, f(z)\big) : z \in \mathcal{H}\},$$

then an isometry f is

(a)   hyperbolic if and only if $\mu_f > 0$;

(b)   elliptic if and only if $\mu_f = 0$ and this infimum is attained;

(a)   parabolic if and only if $\mu_f = 0$ but this infimum is not attained.

5.   Elementary groups of isometries

A group G of isometries of $\mathcal{H}$ is an *elementary group* if there is a nonempty G-invariant set (in the Euclidean closure of $\mathcal{H}$) that contains at most two points. The elementary groups have a simple structure and are easily analysed, and, in general, we are only interested in theorems about non-elementary groups.

If G is an elementary group with an invariant point ζ in $\mathcal{H}$, then G is a group of hyperbolic rotations about ζ (so that G contains only elliptic elements and I). If $\mathcal{H} = \mathbb{D}$ and $\zeta = 0$, then G is a group of Euclidean rotations about 0. If G is an elementary group with an invariant set $\{\zeta_1, \zeta_2\}$, where $\zeta_1, \zeta_2 \in \mathcal{H}$, we may assume that $\mathcal{H} = \mathbb{D}$ and that $\zeta_1 = 0$; we then see that G contains only the identity.

If G is an elementary group with an invariant point ζ in $\partial\mathcal{H}$, then G is a group of parabolic, and hyperbolic, isometries that fix ζ. If $\mathcal{H} = \mathbb{H}$ and $\zeta = \infty$, then G is a group of Euclidean similarities $z \mapsto az + b$, where $a > 0$ and $b \in \mathbb{R}$. If G is an elementary group with an invariant set $\{\zeta_1, \zeta_2\}$, where $\zeta_1 \in \mathcal{H}$ and $\zeta_2 \in \partial\mathcal{H}$, we may assume that $\mathcal{H} = \mathbb{H}$, $\zeta_1 = \infty$ and $\zeta_2 = i$; we then see that G contains only the identity.

If G is an elementary group with an invariant set $\{\zeta_1, \zeta_2\}$, where $\zeta_1, \zeta_2 \in \partial\mathcal{H}$, then G contains only hyperbolic isometries whose axis is the geodesic γ with endpoints ζ_1 and ζ_2, and elliptic isometries, each of which is a rotation of angle π about some point of γ. Indeed, if f is parabolic, then the orbit of ζ_1, or of ζ_2, under powers of f is infinite, so that $f \notin G$.

It is easy to see that *every finite group of isometries is elementary*, for suppose that G is such a group. We may assume that G acts on $\mathbb{D}$, so take any point of $\mathbb{D}$ and let its G-orbit be $\{z_1, \ldots, z_n\}$. There is a unique smallest closed hyperbolic disc that contains this set, so (by the uniqueness), this disc is invariant under every element of G. We deduce that every element of G fixes the centre of this disc so that G is elementary. This argument shows that every non-elementary group of isometries is infinite; the next result says a little more than this.

Theorem 5.1. *If G is a non-elementary group of isometries, then every point in the extended complex plane $\mathbb{C} \cup \{\infty\}$ has an infinite G-orbit.*

Proof   We assume that some point of the extended plane has a finite G-orbit, say $\mathcal{O}$, and we have to show that G is elementary. We may suppose that G acts on $\mathbb{D}$ and, as the action of G respects reflection in the unit circle, we may suppose that either $\mathcal{O} \subset \partial\mathbb{D}$, or $\mathcal{O} \subset \mathbb{D}$.

If $\mathcal{O} \subset \mathbb{D}$ then there is a unique smallest closed hyperbolic disc containing $\mathcal{O}$, and this disc must itself be G-invariant. It follows that the centre of this disc is G-invariant, so that in this case, G is elementary.

If $\mathcal{O} \subset \partial\mathbb{D}$, we distinguish between the cases when $\mathcal{O}$ has exactly three elements, or more than three elements. If $\mathcal{O}$ has exactly three elements, we join them by geodesics and it is clear that there is a unique largest closed hyperbolic disc that lies within this closed triangle. As this disc is G-invariant, the argument above shows that in this case the elements of G have a common fixed point.

Suppose now that $\mathcal{O}$ has at least four elements. In this case we join each pair of points in $\mathcal{O}$ by a geodesic, to obtain a finite G-invariant set of geodesics, some of which intersect others in the set. It follows that the set of intersections points of pairs of these geodesics is a finite G-invariant set of points in $\mathbb{D}$ and, as we have seen above, this implies that G is elementary. The proof is now complete.

Finally, we show that any non-elementary group of isometries must contain parabolic, or hyperbolic, isometries.

Theorem 5.2. *Suppose that a group G of isometries contains only elliptic elements and I. Then G is elementary.*

Proof We shall assume that G is non-elementary and reach a contradiction. Take any nontrivial element g of G; then g is a hyperbolic rotation of some angle θ about its fixed point ζ_1, say. As G is non-elementary, it contains some element h which does not fix ζ_1; then hgh^{-1} is a hyperbolic rotation in G of angle θ about its fixed point ζ_2, where $\zeta_2 = h(\zeta_1)$ and $\zeta_1 \neq \zeta_2$.

Now let γ be the geodesic through ζ_1 and ζ_2. Also, let α be a geodesic through ζ_1 making an angle $\theta/2$ with γ, and let β be a geodesic through ζ_2 making an angle $\theta/2$ with γ. Elementary geometry shows that α and β do not intersect in $\mathbb{D}$ so that the composition $R_\alpha R_\beta$ is not elliptic (or I). However,

$$R_\alpha R_\beta = \big(R_\alpha \circ R_\gamma\big) \circ \big(R_\gamma \circ R_\beta\big),$$

and as the first element on the right is g or g^{-1}, and the second element is hgh^{-1} or its inverse, we see that the non-elliptic element $R_\alpha R_\beta$ is in G. This is the contradiction we are seeking.

6. Groups without elliptic elements

Our main objective is to prove the following result, and then to apply it to the geometry of Riemann surfaces.

Theorem 6.1. *Suppose that f and g are isometries of $\mathcal{H}$ that generate a non-elementary group G. If G has no elliptic isometries then, for all z in $\mathcal{H}$,*

$$\sinh \tfrac{1}{2}\rho\big(z, f(z)\big) \sinh \tfrac{1}{2}\rho\big(z, g(z)\big) \geq 1. \tag{6.1}$$

This result shows that for any z in $\mathcal{H}$,

$$\max\{\sinh \tfrac{1}{2}\rho\big(z, f(z)\big), \sinh \tfrac{1}{2}\rho\big(z, g(z)\big)\} \geq 1,$$

and this, in turn, shows that *every point z of $\mathcal{H}$ is moved a hyperbolic distance of at least* $2\sinh^{-1}(1)$ *$(= 1.762\cdots)$ by f or by g.* Notice that we are *not* assuming that the group G in Theorem 6.1 is discrete (a definition of discreteness will be given later) and it is a *corollary* of Theorem 6.1 that *if G is a non-elementary group of isometries without elliptic elements, then G is discrete* (we shall prove this in Section 12).

For another application of Theorem 6.1, suppose that f and g are isometries of $\mathbb{D}$ and that $\langle f, g \rangle$ is non-elementary and has no elliptic elements. Then

$$1 \leq \sinh^2 \tfrac{1}{2}\rho(0, f(0)) \sinh^2 \tfrac{1}{2}\rho(0, g(0)) = \left(\frac{|f(0)|^2}{1 - |f(0)|^2} \right) \left(\frac{|g(0)|^2}{1 - |g(0)|^2} \right),$$

from which we deduce that

$$|f(0)|^2 + |g(0)|^2 \geq 1. \tag{6.2}$$

It follows that under these circumstances, *either f or g moves the origin 0 a Euclidean distance of at least* $1/\sqrt{2}$ $(= 0.707\cdots)$.

The proof of Theorem 6.1 consists of

(1)   showing that the possibility that f and g are hyperbolic with parallel axes does not arise (see Section 7);

and then establishing the inequality (6.1) in each of the following cases :

(2) f and g are hyperbolic with disjoint axes (see Section 8);

(3) f and g are hyperbolic with crossing axes (see Section 9);

(4) f and g are parabolic (see Section 10);

(5) f is hyperbolic and g is parabolic (see Section 11).

Finally, we remark that as all of the concepts in Theorem 6.1, and the terms in the inequality (6.1), are invariant under conjugation, we may replace f and g by conjugate elements hfh^{-1} and hgh^{-1} (for any suitable h) in any part of our proof of Theorem 6.1.

7. Two hyperbolic isometries with parallel axes

The only result in this section is as follows, and this is (1) in our proof of Theorem 6.1.

Lemma 7.1. *Suppose that G is a non-elementary group of isometries of $\mathcal{H}$, and that G contains two hyperbolic isometries f and g whose axes have exactly one endpoint in common. Then G contains elliptic isometries.*

Proof It is possible to give a geometric proof of this, but an algebraic argument is shorter. By considering a conjugate group we may assume that $\mathcal{H} = \mathbb{H}$, that the common endpoint of the axes is ∞, and that f fixes 0. This implies that $f(z) = kz$ and $g(z) = az + b$, where $k > 0$, $k \neq 1$ and $a > 0$, $a \neq 1$ and $b \in \mathbb{R}$. By replacing f with f^{-1} (if necessary), we can also assume that $k > 1$.

Let $t_n = f^{-n}gf^ng^{-1}$; then a calculation shows that

$$t_n(z) = f^{-n}gf^ng^{-1}(z) = z - b + bk^{-n},$$

so that G contains translations τ_n $(= t_{n+1}t_n^{-1})$, where $\tau_n(z) = z + \mu_n$ and where $\mu_n \neq 0$ but $\mu_n \to 0$. As G is non-elementary, it contains some isometry h that does not fix ∞, so we can write $h(z) = (a_1z + b_1)/(c_1z + d_1)$, where $a_1d_1 - b_1c_1 = 1$ and $c_1 \neq 0$. A simple calculation now shows that

$$\text{trace}(\tau_n h\tau_n h^{-1}) = 2 - (\mu_n c_1)^2,$$

and as this lies in the open interval $(1, 2)$ for all sufficiently large n (because $c_1 \in \mathbb{R}$ and $\mu_n \to 0$), the isometry $\tau_n h\tau_n h^{-1}$ is elliptic for these n.

8. Two hyperbolic isometries with disjoint axes

Throughout this section we shall assume

(1) f and g are hyperbolic isometries with axes $\mathcal{A}_f, \mathcal{A}_g$ and translation lengths T_f, T_y;

(2) $\mathcal{A}_f$ and $\mathcal{A}_g$ are disjoint and have no common endpoints;

(3) the group $\langle f, g \rangle$ has no elliptic elements.

Our first result (Lemma 8.1) shows that given T_f and T_g, the axes $\mathcal{A}_f$ and $\mathcal{A}_g$ cannot be too close together. This implies that no point z can be close to both axes, and the second result (Lemma 8.2) gives a precise bound on how close z can be to the pair of axes.

Lemma 8.1. *Suppose that (1), (2) and (3) hold. Then*

$$\sinh \tfrac{1}{2}T_f \sinh \tfrac{1}{2}T_g \sinh^2 \tfrac{1}{2}\rho(\mathcal{A}_f, \mathcal{A}_g) \geq 1. \tag{8.1}$$

Lemma 8.2. *Suppose that* (1), (2) *and* (3) *hold. Then, for any* z *in* $\mathcal{H}$,

$$\cosh \rho(z, \mathcal{A}_f) \cosh \rho(z, \mathcal{A}_g) \geq \cosh^2 \tfrac{1}{2} \rho(\mathcal{A}_f, \mathcal{A}_g). \tag{8.2}$$

If we now combine (4.2), (8.1) and (8.2), we immediately obtain the following main result of this section which is also (2) in our proof of Theorem 6.1.

Theorem 8.3. *Suppose that* (1), (2) *and* (3) *hold. Then for all* z *in* $\mathbb{D}$,

$$\sinh \tfrac{1}{2} \rho(z, fz) \sinh \tfrac{1}{2} \rho(z, gz) \geq \frac{1}{\tanh^2 \tfrac{1}{2} \rho(\mathcal{A}_f, \mathcal{A}_g)} \geq 1. \tag{8.3}$$

We have only to prove Lemmas 8.1 and 8.2.

The proof of Lemma 8.1   We consider a conjugate group such that the common orthogonal segment to $\mathcal{A}_f$ and $\mathcal{A}_g$ lies on the imaginary axis $\mathbb{I}$, and has midpoint at the origin. Let γ_f and γ_g be the geodesics illustrated in Figure 8.1; for example, γ_f is orthogonal to $\mathcal{A}_f$ and cuts $\mathcal{A}_f$ at a distance $\tfrac{1}{2}T_f$ from $\mathbb{I}$, and γ_g is determined similarly and lies on the same side of $\mathbb{I}$ as γ_f. Now let σ, σ_f and σ_g denote reflections in $\mathbb{I}$, γ_f and γ_g, respectively. Then f (or possibly f^{-1}) is $\sigma_f \sigma$, and g (or g^{-1}) is $\sigma \sigma_g$. Because $\sigma^2 = I$, it follows that $\sigma_f \sigma_g$ is in G and hence is not elliptic, and this, in turn, means that γ_f and γ_g cannot intersect in $\mathbb{D}$. It follows that we can construct a geodesic γ^* (as in Figure 8.1) that separates γ_f from γ_g. Of course, the geodesics γ_f, γ^* and γ_g may have a common endpoint (this is the situation in the extreme case), but certainly γ_f and γ_g cannot intersect in $\mathbb{D}$.

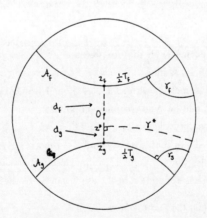

Figure 8.1

Suppose that $\mathcal{A}_f$, $\mathcal{A}_g$ and γ^* meet $\mathbb{I}$ at z_f, z_g and z^*, respectively, and let $d_f = \rho(z^*, z_f)$, and $d_g = \rho(z^*, z_g)$; see Figure 8.1. From a classical formula in hyperbolic trigonometry, we have

$$\sinh \tfrac{1}{2} T_f \sinh d_f \geq 1, \quad \sinh \tfrac{1}{2} T_g \sinh d_g \geq 1.$$

Next, let D and d be defined by $2D = d_f + d_g$ and $2d = d_f - d_g$, where we may assume (as in Figure 8.1) that $d_f \geq d_g$. Then

$$\begin{aligned}
\sinh d_f \sinh d_g &= \sinh(D + d) \sinh(D - d) \\
&= \sinh^2 D - \sinh^2 d \\
&\leq \sinh^2 D \\
&= \sinh^2 \tfrac{1}{2} \rho(\mathcal{A}_f, \mathcal{A}_g),
\end{aligned}$$

and the inequality (8.1) follows immediately.

The proof of Lemma 8.2 Let $\Phi(z) = \cosh \rho(z, \mathcal{A}_f) \cosh \rho(z, \mathcal{A}_g)$, and $\mu = \inf_z \Phi(z)$. It is clear that Φ takes the value μ at some point (for $\Phi(z) \to +\infty$ as $|z| \to 1$), and it is also clear that μ is attained only at a point that lies on or between the two axes $\mathcal{A}_f$ and $\mathcal{A}_g$. Also, another simple argument of this type shows that if $\Phi(z) = \mu$, then z lies on the segment orthogonal to both $\mathcal{A}_f$ and $\mathcal{A}_g$.

Now suppose that $\Phi(z_0) = \mu$, so that (in Figure 8.1) z_0 lies on the segment $[z_g, z_f]$ of $\mathbb{I}$ that is between $\mathcal{A}_f$ and $\mathcal{A}_g$. Let $\delta_f = \rho(z_0, z_f)$ and $\delta_g = \rho(z_0, z_g)$, and, as before, write $\Delta = \delta_f + \delta_g$ and $\delta = \delta_f - \delta_g$. Then, for all z,

$$\begin{aligned}
\cosh \rho(z, \mathcal{A}_f) \cosh \rho(z, \mathcal{A}_g) &\geq \cosh \rho(z_0, \mathcal{A}_f) \cosh \rho(z_0, \mathcal{A}_g) \\
&= \cosh \delta_f \cosh \delta_g \\
&= \cosh(\Delta + \delta) \cosh(\Delta - \delta) \\
&= \cosh^2 \Delta + \sinh^2 \delta \\
&\geq \cosh^2 \Delta \\
&= \cosh^2 \tfrac{1}{2} \rho(\mathcal{A}_f, \mathcal{A}_g)
\end{aligned}$$

and this is (8.2).

9. Two hyperbolic isometries with crossing axes

Again, we assume that f and g are hyperbolic with axes $\mathcal{A}_f, \mathcal{A}_g$ and translation lengths

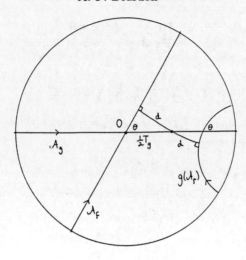

Figure 9.1

T_f, T_g, respectively, and that $\langle f, g \rangle$ has no elliptic elements. We also assume that $\mathcal{A}_f$ and $\mathcal{A}_g$ intersect at an angle θ, and by replacing either or both of f and g by their inverses, we may assume that the situation is as illustrated in Figure 9.1, where $0 < \theta < \pi$.

Let $h = gfg^{-1}$; then (by elementary hyperbolic trigonometry) we can see that $\mathcal{A}_g$ meets the common orthogonal of $\mathcal{A}_f$ and $\mathcal{A}_h$ (which is $g(\mathcal{A}_f)$) at its midpoint. Thus, with d as in Figure 9.1, $\rho(\mathcal{A}_f, \mathcal{A}_h) = 2d$ and

$$\sinh d = \sinh \tfrac{1}{2} T_g \sin \theta \le \sinh \tfrac{1}{2} T_g. \tag{9.1}$$

Next, Lemma 8.1 (applied to f and h) gives

$$\sinh \tfrac{1}{2} T_f \sinh \tfrac{1}{2} T_h \sinh^2 d \ge 1, \tag{9.2}$$

and as $T_h = T_f$, (9.1) and (9.2) give $\sinh \tfrac{1}{2} T_f \sinh \tfrac{1}{2} T_g \ge 1$. As $\sinh \tfrac{1}{2} \rho(z, fz) \ge \sinh \tfrac{1}{2} T_f$, and similarly for g, we deduce that for all z,

$$\sinh \tfrac{1}{2} \rho(z, fz) \sinh \tfrac{1}{2} \rho(z, gz) \ge 1,$$

and this is (3) in our proof of Theorem 6.1.

10.   Two parabolic isometries

Suppose that f and g are parabolic isometries, and that they generate a non-elementary group. As this group is non-elementary, f and g must have distinct fixed points. In fact the converse is true (if f and g have distinct fixed points then they generate a non-elementary group) but we shall not need this. In order to complete this part of our proof of Theorem 6.1, we need to prove the following lemma.

Lemma 10.1. *Suppose that f and g are parabolic isometries of $\mathcal{H}$ with distinct fixed points, and that the group G generated by them contains no elliptic elements. Then, for all z in $\mathcal{H}$,*

$$\sinh \tfrac{1}{2}\rho\big(z, f(z)\big) \sinh \tfrac{1}{2}\rho\big(z, g(z)\big) \geq 1. \tag{10.1}$$

We shall give two proofs of this. By considering a conjugate group, and by replacing f by f^{-1}, and g by g^{-1}, as necessary, we may assume that $\mathcal{H} = \mathbb{H}$, and

$$f(z) = z + 1, \quad g(z) = \frac{z}{az + 1},$$

where $a > 0$.

The first proof The formula (2.6) shows that

$$\sinh \tfrac{1}{2}\rho\big(z, f(z)\big) \sinh \tfrac{1}{2}\rho\big(z, g(z)\big) = \frac{|z - f(z)|.|z - g(z)|}{4y(\operatorname{Im}[f(z)]\operatorname{Im}[g(z)])^{1/2}},$$

and a little calculation shows that the right hand side of this is $a|z|^2/4y^2$. As $|z| \geq y$, it only remains to show that $a \geq 4$.

Let γ be the geodesic given by $x = 0$; let α be the geodesic given by $x = 1/2$; let β be the geodesic given by $|z - 1/a| = 1/a$. It is clear that $f = R_\alpha \circ R_\gamma$ (R_α is the reflection in α, and so on), and it is known (and easy to see) that $g^{-1} = R_\gamma \circ R_\beta$ (see Figure 10.1). We deduce that $R_\alpha \circ R_\beta \in G$, and as G has no elliptic elements, this means that α and β do not meet in $\mathbb{H}$. We deduce that $2/a \leq 1/2$; thus $a \geq 4$ and this completes the first proof.

The second proof is based on the idea of the canonical horocycle for a parabolic isometry; see Section 4.

The second proof The inequality (10.1) is an easy consequence of the fact that the canonical horocycles H_f and H_g are exterior (but possibly tangent) to each other, for suppose that this is true. Then H_f and H_g divide $\mathcal{H}$ into three regions, namely

(a) the set of points inside H_f;

(b) the set of points inside H_g, and

(c) the set of points exterior to both H_f and H_g.

If z is exterior to both H_f and H_g then, by Lemma 4.1,

$$\sinh \tfrac{1}{2}\rho\big(z, f(z)\big) \geq 1, \quad \sinh \tfrac{1}{2}\rho\big(z, g(z)\big) \geq 1,$$

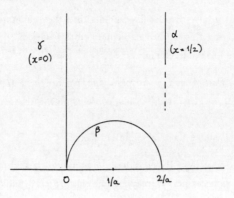

Figure 10.1

so that (10.1) holds in this case. If z lies inside H_f, and hence outside H_g, then $\rho(z, H_f) \le$ $\rho(z, H_g)$ so that, by Lemma 4.1,

$$\sinh \tfrac{1}{2}\rho\big(z, f(z)\big) \sinh \tfrac{1}{2}\rho\big(z, g(z)\big) = \exp\left[\rho(z, H_g) - \rho(z, H_f)\right] \ge 1$$

as required. The case when z lies inside U_g is similar.

It remains to prove that H_f and H_g are exterior (but perhaps tangent) to each other, and this is a direct consequence of the fact that the geodesics α and β in the first proof do not meet in $\mathbb{H}$. Indeed, with f and g as in the first proof, we have $H_f = \{x + iy : y = 1/2\}$ and

$$H_g = \{z \in \mathbb{H} : |z - i/a| = 1/a\} \subset \{x + iy : y \le 2/a\} \subset \{x + iy : y \le 1/2\}.$$

11. A parabolic and a hyperbolic isometry

In this section we consider case (5) in the proof of Theorem 6.1, and we shall assume that f is a hyperbolic isometry and that g is a parabolic isometry. As f and g generate a non-elementary group, they have no common fixed point; thus, by considering a conjugate group, and by replacing one or both of f and g by their inverses if necessary, we may assume that $\mathcal{H} = \mathbb{H}$, $f(z) = z + a$ where $a > 0$, and g has fixed points b and $-b$, where $b > 0$.

Now let γ be the geodesic given by $x = 0$; let α be the geodesic given by $x = a/2$; let β be the geodesic orthogonal to A_g so that the distance between γ and β (measured along A_g) is $\tfrac{1}{2}T_g$: see Figure 11.1.

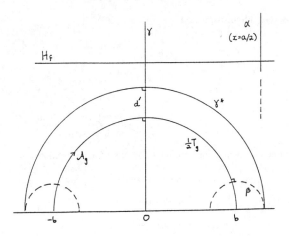

Figure 11.1

It is clear that f (or f^{-1}) is $= R_\alpha \circ R_\gamma$, and that g (or g^{-1}) is $R_\gamma \circ R_\beta$ (see Figure 11.1). We deduce that $R_\alpha \circ R_\beta \in G$, and as G has no elliptic elements, this means that α and β do not meet in $\mathbb{H}$. This justifies the construction of the geodesic γ^* which separates α and β as in Figure 11.1, and which has a common endpoint with β. As α lies to the right of γ^* (except possibly with a common endpoint), it follows that γ^* lies underneath the canonical horocycle H_f of f (because H_f is given by $y = a/2$).

We need a preliminary lemma which is actually an immediate corollary of a classical result from hyperbolic geometry.

Lemma 11.1. *Let d be the distance between A_g and H_f (measured along γ). Then*

$$\sinh d \, \sinh(\tfrac{1}{2}T_g) \geq 1.$$

Proof Consider the quadilateral bounded by arcs of γ^*, γ, A_g and β. This quadilateral has angles $\pi/2$, $\pi/2$, $\pi/2$ and 0, and it has sides of lengths d', where $d' \leq d$, $\frac{1}{2}T_g$, $+\infty$ and ∞. It is a classical fact that in such a polygon,

$$\sinh d' \, \sinh(\tfrac{1}{2}T_g) \geq 1,$$

and the desired result follows because $d > d'$.

It is now easy to complete the proof of the inequality (6.1) in this case. Suppose first that

z lies on or above H_f. Then, from Lemma 4.1 and (4.2),

$$\sinh \tfrac{1}{2}\rho\big(z, f(z)\big) \sinh \tfrac{1}{2}\rho\big(z, g(z)\big) = \frac{\sinh(\tfrac{1}{2}T_g) \cosh \rho(z, A_g)}{\exp \rho(z, H_f)}$$

$$\geq \frac{\cosh \rho(z, A_g)}{\sinh d \, \exp \rho(z, H_f)}$$

$$\geq \frac{\cosh[\rho(z, H_f) + d]}{\sinh d \, \exp \rho(z, H_f)},$$

the second line following because of Lemma 11.1. Now for any real u and v, we have

$$\cosh(u + v) = \cosh u \, \cosh v + \sinh u \, \sinh v$$

$$\geq \cosh u \, \sinh v + \sinh u \, \sinh v$$

$$= \exp(u) \sinh v,$$

and this with the inequality above shows that when z is on or above H_f,

$$\sinh \tfrac{1}{2}\rho\big(z, f(z)\big) \sinh \tfrac{1}{2}\rho\big(z, g(z)\big) \geq 1. \tag{11.1}$$

There remains the case when z is below H_f and this is similar. In this case Lemmas 4.1 and 11.1 yield

$$\sinh \tfrac{1}{2}\rho\big(z, f(z)\big) \sinh \tfrac{1}{2}\rho\big(z, g(z)\big) = \sinh(\tfrac{1}{2}T_g) \cosh \rho(z, A_g) \exp \rho(z, H_f)$$

$$\geq \frac{\cosh \rho(z, A_g) \exp \rho(z, H_f)}{\sinh d}.$$

Now

$$\rho(z, H_f) + \rho(z, A_g) \geq \rho(H_f, A_g) = d,$$

and if u, v and w are real with $u + v \geq w$, then

$$\exp(u) \cosh v = \cosh u \, \cosh v + \sinh u \, \cosh v$$

$$\geq \sinh(u + v)$$

$$\geq \sinh w,$$

so again (11.1) follows. This completes the proof in case (5) of Theorem 6.1.

12.   Fuchsian groups and Riemann surfaces

A *Fuchsian group* G is a discrete group of isometries of the hyperbolic plane $\mathcal{H}$. We have to describe what we mean by 'discrete' here, and this is precisely what is needed in order to make the topological quotient space $\mathcal{H}/G$ a reasonable topological space.

There are several equivalent definitions of a Fuchsian group, and we shall describe these without proving their equivalence. First, there is the *geometric definition*: a group G of isometries acting on $\mathcal{H}$ is Fuchsian if one (and hence every) orbit in $\mathcal{H}$ accumulates only at points of $\partial\mathcal{H}$. The fact that if this holds for one orbit, then it holds for every orbit is a direct consequence of the fact that the group elements are isometries.

If we regard the isometries as matrices (or if we prefer to study groups of 2×2 matrices with determinant 1), we can use the matrix norm $\|A\|$ given by

$$\|A\|^2 = |a|^2 + |b|^2 + |c|^2 + |d|^2, \quad A = \begin{pmatrix} a & b \\ c & d \end{pmatrix},$$

to impose the natural metric on the space of matrices. The group G is then Fuchsian, or discrete, if it is a discrete topological space in its own right (that is, if every point is an isolated point). The link between these two ideas is the beautiful formula (given here when $\mathcal{H} = \mathbb{D}$)

$$\|g\|^2 = 2\cosh\rho\big(0, g(0)\big), \tag{12.1}$$

where

$$\|g\|^2 = |a|^2 + |b|^2 + |c|^2 + |d|^2, \quad g(z) = \frac{az+b}{cz+d}, \quad ad - bc = 1.$$

To get the corresponding formula for $\mathbb{H}$ we simply replace 0 by i. It follows from (12.1) that the orbit of 0 accumulates only on $\partial\mathbb{D}$ if and only if the numbers $\|g\|$, for g in G, accumulate only at $+\infty$, and this is so if and only if the group G is discrete in the sense of the second definition above.

Finally, we can also say that the group G is discrete if (as a topological group) the identity I is an isolated element of G. This means that if $g_1, g_2, \ldots$ are in G and if $g_n \to I$ (in the topology of the group) then $g_n = I$ for all sufficiently large n.

It is a direct consequence of (12.1) that a group of isometries of $\mathcal{H}$ is discrete (as a topological group) if and only if it *acts properly discontinuously in* $\mathcal{H}$, and this means that for any compact subset K of $\mathcal{H}$ the set

$$\{g \in G : g(K) \cap K \neq \emptyset\}$$

is finite. If either of these conditions holds, then G is a Fuchsian group. Note that by taking K to be the singleton set $\{w\}$, we see that given a Fuchsian group G, the stabiliser $\{g \in G : g(w) = w\}$ of w in $\mathcal{H}$ is finite. In particular, every elliptic element in a Fuchsian group must have finite order.

We can now prove the following result.

Theorem 12.1. *Suppose that G is a non-elementary group of isometries of $\mathcal{H}$, and that G does not contain any elliptic elements. Then G is a Fuchsian group.*

The following lemma will be used in our proof of Theorem 12.1; in this, $\mathrm{Fix}(f)$ denotes the set of fixed points (in the extended complex plane) of a Möbius map f.

Lemma 12.2. *Suppose that f and g are isometries of $\mathcal{H}$ (neither the identity), and that the group $\langle f, g \rangle$ generated by them contains no elliptic elements. Then $\langle f, g \rangle$ is elementary if and only if $\mathrm{Fix}(f) = \mathrm{Fix}(g)$.*

Proof Let $G = \langle f, g \rangle$. Clearly, if $\mathrm{Fix}(f) = \mathrm{Fix}(g)$, then G is elementary for the set $\mathrm{Fix}(f)$ is a nonempty G-invariant set with at most two points. We suppose now that G is elementary, and consider the three cases (a) f and g parabolic, (b) f and g hyperbolic, and (c) f, say, is parabolic and g is hyperbolic. We need to show that in each case, $\mathrm{Fix}(f) = \mathrm{Fix}(g)$.

Case (a) As f is parabolic it has a unique fixed point ζ, say, so that $\mathrm{Fix}(f) = \{\zeta\}$. As every point of the extended plane except ζ has an infinite orbit under powers of f, the nonempty finite G-invariant set that exists because G is elementary must be $\{\zeta\}$. As this set is invariant under g, we see that g fixes ζ (and nothing else because g is parabolic). It follows that $\mathrm{Fix}(g) = \{\zeta\}$, and the proof of Case (a) is complete.

Case (b) In this case, both f and g are hyperbolic. Notice that in this case every point z outside $\mathrm{Fix}(f)$ has an infinite G-orbit (consider $f^n(z)$ for $n = 1, 2, \ldots$), and similarly for g. Thus, as G is elementary, we see that $\mathrm{Fix}(f) \cap \mathrm{Fix}(g)$ is nonempty. By Lemma 7.1, this intersection cannot contain exactly one point (for if it does, then $\langle f, g \rangle$ will contain elliptic elements), so the only remaining alternative is that $\mathrm{Fix}(f) = \mathrm{Fix}(g)$.

Case (c) Finally we assume that f is parabolic and g is hyperbolic, say with $\mathrm{Fix}(f) = \{\zeta\}$, and $\mathrm{Fix}(g) = \{\eta_1, \eta_2\}$. As this means that we cannot have $\mathrm{Fix}(f) = \mathrm{Fix}(g)$, we seek a contradiction (which then shows that this case cannot arise).

Exactly as in case (a), we see that $\{\zeta\}$ is G-invariant, and this means that g fixes ζ; thus $\zeta = \eta_1$, say, so that f fixes η_1. The two hyperbolic elements g and fgf^{-1} are in G, and they have fixed point sets $\{\eta_1, \eta_2\}$ and $\{\eta_1, f(\eta_2)\}$, respectively. As $f(\eta_2) \neq \eta_2$, Lemma 7.1 is applicable and this shows that the group generated by g and fgf^{-1} contains elliptic elements. It follows that G contains elliptic elements and this is the contradiction we are seeking. The proof of Lemma 12.2 is now complete.

The proof of Theorem 12.1 We suppose that G is non-elementary, has no elliptics, and is not discrete, and we need to reach a contradiction. Without los of generality we may assume that $\mathcal{H} = \mathbb{D}$. As G is not discrete, it contains elements h_n, where $h_n \to I$, where I is the identity. This implies that $h_n(0) \to 0$ as $n \to \infty$, so that by discarding some of the h_n, we may assume that for all n, $|h_n(0)| < \frac{1}{2}$. It follows that for all m and n,

$$|h_m(0)|^2 + |h_n(0)|^2 < 1,$$

so that from (6.2), the group generated by h_m and h_n is elementary. By considering all possible m and n here, we deduce (from Lemma 12.2) that

$$\mathrm{Fix}(h_1) = \mathrm{Fix}(h_2) = \mathrm{Fix}(h_3) = \cdots = F,$$

say. As G is not elementary, there must be some v in G which does not leave F invariant. Thus $\mathrm{Fix}(v) \neq \mathrm{Fix}(h_n)$, and so

$$\mathrm{Fix}(vh_n v^{-1}) = v\big(\mathrm{Fix}(h_n)\big) \neq v\big(\mathrm{Fix}(v)\big) = \mathrm{Fix}(v).$$

This shows that for each n, the group $\langle v, vh_n v^{-1} \rangle$ is non-elementary; hence for all n

$$\sinh \tfrac{1}{2}\rho(0, v(0)) \sinh \tfrac{1}{2}\rho(0, vh_n v^{-1}(0)) \geq 1.$$

This is false, however, for $h_n \to I$, so that $vh_n v^{-1} \to I$, and so $vh_n v^{-1}(0) \to 0$ as $n \to \infty$. This completes the proof of Theorem 12.1.

From now on our primary concern is Riemann surfaces (rather than Fuchsian groups), and it is a consequence of the Uniformisation Theorem that we need only consider groups without elliptic elements. To simplify the results that follow, we shall also exclude parabolic elements from G although the theory we have developed in this paper is sufficient to make similar (but suitably modified) claims for groups with parabolic elements.

Let $\mathcal{R}$ be any hyperbolic Riemann surface, let G be the associated Fuchsian group acting (without elliptic elements) on $\mathbb{D}$, and let $q : \mathbb{D} \to \mathcal{R}$ be the natural quotient map; thus $\mathcal{R}$ is (conformally equivalent to) $\mathbb{D}/G$. Then every curve C on $\mathcal{R}$ lifts (from any admissible starting point) to a curve C' in $\mathbb{D}$. A closed curve C on $\mathcal{R}$ lifts to a curve C' from, say, w to $g(w)$, in $\mathbb{D}$, where $w \in \mathbb{D}$ and $g \in G$. We can deform the curve C' to a geodesic segment (with the same endpoints as C') and this deformation projects to a deformation of C on $\mathcal{R}$ that is a closed curve which is locally a geodesic except possibly at the starting point. More generally, a closed curve on $\mathcal{R}$ is freely homotopic to either a closed geodesic loop (the shortest curve in its free homotopy class) or to a point (this happens when, for example, C is a simple loop about a 'puncture' on $\mathcal{R}$).

A 'puncture' on $\mathcal{R}$ corresponds under q to the fixed point ζ of some parabolic isometry in G, and a loop around the puncture corresponds to the projection of a curve from z to $f(z)$, where f is the parabolic element with this fixed point. For example, suppose that $\mathcal{H} = \mathbb{H}$, and that f is a parabolic element in G with fixed point ∞, such that f generates the stabiliser of ∞. Then, if y is very large, the projection of the horizontal segment from $x + iy$ to $f(x + iy)$ is a very short curve on $\mathcal{R}$ that surrounds the puncture $q(\infty)$ once.

We end with a few comments on the geometry of hyperbolic Riemann surfaces, and we state once again that when suitably modified, these results remain valid for groups with parabolic elements. First, the comments above are enough to show that the claim (1) made in Section 1 is a direct consequence of Theorem 6.1. In fact, we have the following result.

Theorem 12.3. *Let $\mathcal{R}$ be any hyperbolic Riemann surface, so that $\mathcal{R}$ is of the form $\mathbb{D}/G$, where G is a Fuchsian group without elliptic and parabolic elements. Suppose that σ_1 and σ_2, of hyperbolic lengths ℓ_1 and ℓ_2, respectively, are two intersecting closed loops on $\mathcal{R}$. If neither σ_j is homotopic to a power of the other, then*

$$\sinh \tfrac{1}{2}\ell_1 \, \sinh \tfrac{1}{2}\ell_2 \geq 1 \qquad\qquad (12.2)$$

Briefly, we can deform the curves σ_j so that they become the shortest loops in their respective homotopy classes, and it is clearly sufficient to establish the inequality in this case. It now follows that the curves σ_j lift to segments $[z, g(z)]$ and $[z, h(z)]$ lying on the axes of the hyperbolic elements g and h in G, and from this we see that $\ell_1 = T_g$ and $\ell_2 = T_h$; thus (12.2) is equivalent to the inequality

$$\sinh(\tfrac{1}{2}T_g) \, \sinh(\tfrac{1}{2}T_h) \geq 1. \qquad\qquad (12.3)$$

The case when σ_i is homotopic to a power of σ_j corresponds to the case when the group generated by g and h is elementary (that is, when g and h have the same axis). As we are excluding this case, we can apply Theorem 6.1 and the inequality (6.1) in this certainly implies (12.3).

Finally, we consider the claim (2) in Section 1; this is as follows.

Theorem 12.4. *Let $\mathcal{R}$ be any hyperbolic Riemann surface, and suppose that σ, of hyperbolic length ℓ, is a closed loop on $\mathcal{R}$ that crosses itself at an angle θ, where $0 < \theta < \pi$. Then $\sinh \tfrac{1}{2}\ell \geq 1$.*

Proof   We may deform σ in its free homotopy class and so assume that it is a geodesic loop that lifts to an axis of some hyperbolic element g in $\mathcal{H}$. In fact, this assertion can be justified as follows: lift σ to a curve σ' from z to $g(z)$, say, in $\mathcal{H}$. Now g is hyperbolic (because we have excluded elliptic and parabolic elements from our group); we can therefore deform the curve σ' to a segment $[w, g(w)]$, where w lies on the axis $\mathcal{A}_g$ of g, and this deformation is equivalent to the given deformation on $\mathcal{R}$. Moreover, we see from this argument that $\ell \geq T_g$.

The fact that σ crosses itself on $\mathcal{R}$ means that in $\mathcal{H}$, the axis $\mathcal{A}_g$ crosses some image, say $h(\mathcal{A}_g)$, of itself. We saw in Section 9 that if f and g are hyperbolic elements in a non-elementary group without elliptic elements, and if the axes of f and g cross, then

$$\sinh \tfrac{1}{2}T_f \sinh \tfrac{1}{2}T_g \geq 1.$$

We can apply this with $f = hgh^{-1}$; clearly f is hyperbolic, its axis is $h(\mathcal{A}_g)$, and $T_f = T_g$. We deduce that $(\sinh \tfrac{1}{2}\ell)^2 \geq (\sinh \tfrac{1}{2}T_g)^2 \geq 1$ as required, and this completes our discussion.

The author thanks Catherine Kelk and Raquel Águeda-Maté for their careful reading of an earlier version of this manuscript. The reader may wish to consult the following texts on this subject.

Bibliography

1. Ahlfors, L.V., *Conformal Invariants*, McGraw-Hill, 1973.

2. Beardon, A.F., *The geometry of discrete groups*, Springer-Verlag, 1983.

3. Beardon, A.F., *A Primer on Riemann Surfaces*, London Math. Society Lecture Notes 78, Cambridge Univ. Press, 1984.

4. Buser, P., *Geometry and Spectra of Compact Riemann Surfaces*, Birkhäuser, 1992.

5. Carathéodory, C., *Conformal Representation*, Cambridge Tracts No. 28, Second Edition, Cambridge Univ. Press, 1952.

6. Fenchel, W., *Elementary Geometry in Hyperbolic Space*, de Gruyter, 1989.

7. Ford, L.R., *Automorphic functions*, Second Edition, Chelsea, 1951.

8. Gilman, J., *Two-Generator Discrete Subgroups of* $PSL(2, \mathbb{R})$, Memoirs of American Math. Soc, 561, 1995.

9. Jones, G.A. and Singerman, D., *Complex functions*, Cambridge Univ. Press, 1987.

10. Lehner, J., *Discontinuous Groups and Automorphic Functions*, Mathematical Surveys VIII, American Math. Soc., 1964.

11. Lehner, J., *A Short Course on Automorphic Functions*, Holt, Rinehart and Winston, 1966.

12. Magnus, W., *Noneuclidean Tesselations and their Groups*, Academic Press, 1974.

13. Springer, G., *Introduction to Riemann Surfaces*, Addison-Wesley, 1957.

14. Ramsay A. and Richtmyer, R.D., *Introduction to Hyperbolic Geometry*, Universitext, Springer, 1995.

15. Ratcliffe, J.G., *Foundations of Hyperbolic Manifolds*, Graduate Texts 149, Springer-Verlag, 1994.

Department of Pure Mathematics and Mathematical Statistics
University of Cambridge, UK

INTRODUCTION TO ARITHMETIC FUCHSIAN GROUPS

C. Maclachlan

1 Introduction

This article is an extended version of the lecture given at UNED in July 1998 on "Topics on Riemann surfaces and Fuchsian groups" to mark the 25th anniversary of UNED.

The object of that lecture was to motivate the definition of arithmetic Fuchsian groups from the special and very familiar example of the classical modular group. This motivation proceeded via quaternion algebras and the lecture *ended* with the definition of arithmetic Fuchsian groups in these terms. This essay will go a little beyond that to indicate how the number theoretic data defining an arithmetic Fuchsian group can be used to determine geometric and group-theoretic information. No effort is made here to investigate other approaches to arithmetic Fuchsian groups via quadratic forms or to discuss and locate these groups in the general theory of discrete arithmetic subgroups of semi-simple Lie groups. Thus the horizons of this article are limited to giving one method of introducing an audience familiar with the ideas of Fuchsian groups and Riemann surfaces to the interesting special subclass of arithmetic Fuchsian groups.

2 Basics

A Fuchsian group is a *discrete* subgroup of $SL(2,\mathbb{R})$ or of $PSL(2,\mathbb{R}) = SL(2,\mathbf{R})/ < -I >$. We will frequently employ the usual abuse of notation by writing elements of $PSL(2,\mathbb{R})$ as matrices, while strictly they are only determined up to sign. The topology on $SL(2,\mathbb{R})$ and $PSL(2,\mathbb{R})$ is determined by the matrix entries as subset of $\mathbb{R}^4$.

The upper half plane $U = \{z \in \mathbb{C} \mid Im(z) > 0\}$ with the hyperbolic metric defined by $ds = \dfrac{|dz|}{Im\ z}$ is a model of hyperbolic space $\mathbf{H}^2$. The element $\gamma = \begin{pmatrix} a & b \\ c & d \end{pmatrix}$ acts via linear fractional transformation $z \mapsto \dfrac{az+b}{cz+d}$ on U so that $PSL(2,\mathbb{R})$ becomes the full group of orientation-preserving isometries of this model of $\mathbf{H}^2$.

A subgroup Γ is discrete in $(P)SL(2,\mathbb{R})$ if and only if Γ acts discontinuously on U. The group Γ then has a fundamental region in U and we will be mainly concerned with groups which have a fundamental region of finite hyperbolic area. Following standard arithmetic group practice, such groups will be referred to as *lattices* in $PSL(2, \mathbb{R})$.

Notice that, if Γ is a lattice, then other subgroups "close" to it, in the senses described below, are also lattices.

A. Let $\Gamma_1 \subset \Gamma$ as a subgroup of finite index. Then Γ is a lattice if and only if Γ_1 is a lattice.

B. Let $x \in SL(2, \mathbb{R})$. Then Γ is is a lattice if and only if $x\Gamma x^{-1}$ is a lattice.

To cover these, we make use of the following general definitions:

Definition 2.1 *Let Γ_1, Γ_2 be subgroups of G.*

1. *Γ_1 and Γ_2 are commensurable if $\Gamma_1 \cap \Gamma_2$ is of finite index in both Γ_1 and Γ_2.*

2. *Γ_1 and Γ_2 are commensurable in the wide sense if Γ_1 and a conjugate of Γ_2 are commensurable.*

Thus if any subgroup Γ of $SL(2, \mathbb{R})$ in a wide commensurability class is a lattice, then all subgroups in that class are lattices.

Classical Example Let $\Gamma_0 = SL(2, \mathbb{Z})$ - the modular group. Clearly $SL(2, \mathbb{Z})$ is discrete in $SL(2, \mathbb{R})$. Furthermore it is well-known that in its action on U, $PSL(2, \mathbb{Z})$ has a fundamental region of finite hyperbolic area $\pi/3$ and from the geometry of the fundamental region, a presentation of the group as $\mathbb{Z}_2 * \mathbb{Z}_3$ can be obtained.

3 Generalising the Classical Example

First note that

$$SL(2, \mathbb{Z}) = \{X \in M_2(\mathbb{Z}) \mid det(X) = 1\}.$$

Now $M_2(\mathbb{Z})$ is the subring of matrices with integer entries in the algebra $M_2(\mathbb{Q})$. Thus

$$M_2(\mathbb{Z}) \subset M_2(\mathbb{Q}) \quad \text{and} \quad M_2(\mathbb{Z}) \otimes_{\mathbb{Z}} \mathbb{Q} \cong M_2(\mathbb{Q}).$$

Now $M_2(\mathbf{Q})$ is a 4 dimensional matrix algebra over $\mathbb{Q}$ and as such it is central and simple.

Central means that the centre $Z(M_2(\mathbb{Q})) = \mathbb{Q}.I$ i.e. consists of rational multiples of the identity.

Simple means that it has no proper two-sided ideals.

Thus the modular group can be considered as arising from the following chain of constructions:

$$M_2(\mathbb{Q}) \rightsquigarrow M_2(\mathbb{Z}) \rightsquigarrow SL(2,\mathbb{Z}). \tag{1}$$

It is this chain which we generalise; first of all by replacing $\mathbb{Q}$ by any number field k i.e. a finite extension of $\mathbb{Q}$. Then take a 4-dimensional central simple algebra over k. This is a well-studied class of algebras and they have an easily described form.

Theorem 3.1 *Let A be a 4 dimensional central simple algebra over k. Then A has a basis of the form $\{1, i, j, ij\}$ where 1 is a multiplicative identity, $i^2 = a.1, j^2 = b.1$ where $a, b \in k^*$ and $ij = -ji$.*

Definition 3.2 *A 4 dimensional central simple algebra is called a* quaternion *algebra over k and is denoted by a Hilbert symbol* $\left(\dfrac{a,b}{k}\right)$.

For any $a, b \in k^*$, one can define a quaternion algebra over k as in this theorem, so that there are many such quaternion algebras. For quaternion algebras see [19], [12, Chap. 1], [5, Chap. 7], [9, Chap. 3].

Examples

1. $M_2(\mathbf{Q}) = \left(\dfrac{1,1}{\mathbf{Q}}\right)$ where

$$1 = \begin{pmatrix} 1 & 0 \\ 0 & 1 \end{pmatrix}, \quad i = \begin{pmatrix} 1 & 0 \\ 0 & -1 \end{pmatrix}, \quad j = \begin{pmatrix} 0 & 1 \\ 1 & 0 \end{pmatrix}, \quad ij = \begin{pmatrix} 0 & 1 \\ -1 & 0 \end{pmatrix}.$$

2. $M_2(k) = \left(\dfrac{1,1}{k}\right)$. $\left[= \left(\dfrac{1,-1}{k}\right) \text{ by taking } j = \begin{pmatrix} 0 & 1 \\ -1 & 0 \end{pmatrix}. \right]$

3. $\left\{ \begin{pmatrix} a + b\sqrt{2} & 3(c + d\sqrt{2}) \\ c - d\sqrt{2} & a - b\sqrt{2} \end{pmatrix} \mid a, b, c, d \in \mathbb{Q} \right\}$. It can be shown that this is

a 4 dimensional central simple algebra over $\mathbb{Q}$. With $i = \begin{pmatrix} \sqrt{2} & 0 \\ 0 & -\sqrt{2} \end{pmatrix}$,

$j = \begin{pmatrix} 0 & 3 \\ 1 & 0 \end{pmatrix}$ we see that a Hilbert symbol is $\left(\dfrac{2,3}{\mathbb{Q}}\right)$.

I have displayed these examples as subalgebras of matrix algebras, but that is unnecessary as they admit a simple elementwise description via their basis elements and the rules of multiplication given in the theorem above. Thus

4. $\left(\dfrac{-7, e^{2\pi i/5}}{\mathbb{Q}(e^{2\pi i/5})}\right) = \{a_0 + a_1 i + a_2 j + a_3 ij \mid a_i \in \mathbb{Q}(e^{2\pi i/5})\}$ and multiplication defined by $i^2 = -7, j^2 = e^{2\pi i/5}, ij = -ji$.

On any quaternion algebra, there is a conjugate map defined for $\alpha = a_0 + a_1 i + a_2 j + a_3 ij$ by $\bar{\alpha} = a_0 - a_1 i - a_2 j - a_3 ij$. From this, one obtains the (reduced) *norm*, $n(\alpha) = \alpha\bar{\alpha}$ and the (reduced) *trace* $\mathrm{tr}(\alpha) = \alpha + \bar{\alpha}$. On a matrix algebra, the norm is simply the determinant. The conjugate map does not depend on the choice of basis.

It is readily seen, that, in the Hilbert symbol, one can adjoin or remove squares so that $\left(\dfrac{a,b}{k}\right) \cong \left(\dfrac{ax^2, by^2}{k}\right)$. Indeed , the Hilbert symbol is by no means uniquely determined e.g. $\left(\dfrac{a,b}{k}\right) \cong \left(\dfrac{b,ab}{k}\right)$ etc. There is however a powerful existence and classification theorem for quaternion algebras over number fields which will be discussed later.

If we consider quaternion algebras over $\mathbb{R}$, then as we can adjoin or remove squares, there are only two possibilities

$$\left(\frac{1,1}{\mathbb{R}}\right) \cong \left(\frac{1,-1}{\mathbb{R}}\right) \cong M_2(\mathbb{R}) \text{ or } \left(\frac{-1,-1}{\mathbb{R}}\right) \cong \mathcal{H}, \text{Hamilton's quaternions.}$$

Returning to our chain of constructions generalising (1), these quaternion algebras over number fields are the objects which generalise $M_2(\mathbb{Q})$. Note that, filling out that construction gives:

$$
\begin{array}{ccccc}
M_2(\mathbb{Q}) & \rightsquigarrow & M_2(\mathbb{Z}) & \rightsquigarrow & SL(2,\mathbb{Z}) \\
\downarrow \otimes_{\mathbb{Q}}\mathbb{R} & & & & \downarrow \\
M_2(\mathbb{R}) & & & \rightsquigarrow & SL(2,\mathbb{R})
\end{array}
$$

To generalise this then, our field k must be a subfield of the reals and the quaternion algebra A must be such that $A \otimes_k \mathbb{R} \cong M_2(\mathbb{R})$. Thus we choose k to be a real number field, $A = \left(\dfrac{a,b}{k}\right)$ a quaternion algebra over k such that $\left(\dfrac{a,b}{\mathbb{R}}\right) \cong M_2(\mathbb{R})$, which requires that a and b are not both negative.

The intermediate gadget in our derivation is $M_2(\mathbb{Z})$. What is this? It is a ring with 1, a finitely-generated $\mathbb{Z}$-module and a lattice, so that $M_2(\mathbb{Z}) \otimes_{\mathbb{Z}} \mathbb{Q} \cong M_2(\mathbb{Q})$. Thus replacing $\mathbb{Z}$ by the ring of integers R_k in k, we have [13, 19]:

Definition 3.3 *Let A be a quaternion algebra over k and let R_k denote the ring of integers in k. An (R_k)-order $\mathcal{O}$ in A is a subring with 1 which is a finitely generated R_k-module such that $\mathcal{O} \otimes_{R_k} k \cong A$.*

Examples

1. $M_2(R_k)$ in $M_2(k)$.

2. $A = \left(\dfrac{\sqrt{2}, 1+\sqrt{2}}{\mathbb{Q}(\sqrt{2})}\right)$ and $\mathcal{O} = \mathbb{Z}[\sqrt{2}][1, i, j, ij]$.

The last step in our derivation is to obtain $SL(2, \mathbb{Z})$ from $M_2(\mathbb{Z})$ i.e. those of determinant 1. The norm in a quaternion algebra generalises the determinant, so define

$$\mathcal{O}^1 = \{\alpha \in \mathcal{O} \mid n(\alpha) = 1\}.$$

We thus obtain the chain:

$$\mathcal{O}^1 \subset \mathcal{O} \subset A \rightarrow A \otimes_k \mathbb{R} \cong M_2(\mathbb{R})$$

and so a mapping

$$\phi : \mathcal{O}^1 \rightarrow SL(2, \mathbb{R}).$$

4   Arithmetic lattices

BUT, BUT, BUT ...... there is no obvious reason why the image of $\mathcal{O}^1$ under ϕ should be discrete in $SL(2, \mathbb{R})$ and it may not be.

Example Let $\mathbb{Q} \subset k \subset \mathbb{R}$ with $k \neq \mathbb{Q}$. Then

$$SL(2, R_k) \subset M_2(R_k) \subset M_2(k) \rightarrow M_2(k) \otimes_k \mathbb{R} \cong M_2(\mathbb{R})$$

yielding $SL(2, R_k) \subset SL(2, \mathbb{R})$. This is never discrete.

Theorem 4.1 *Let α be a real irrational. Then $\exists$ a sequence $p_n/q_n \in \mathbb{Q}$ such that $\mid \alpha - \dfrac{p_n}{q_n} \mid < \dfrac{1}{q_n^2}$.*

Proof: Take the continued fraction expansion of α, $\alpha = [a_0; a_1, a_2, \dots]$. Since α is irrational, this is an infinite expansion. Let p_n/q_n be the n'th convergent i.e. $p_n/q_n = [a_0; a_1, \dots, a_n]$. Then it can be shown that the inequality holds.

Let $\alpha \in R_k \setminus \mathbb{Z}$. Then the elements

$$\begin{pmatrix} 1 & q_n\alpha - p_n \\ 0 & 1 \end{pmatrix} \rightarrow \begin{pmatrix} 1 & 0 \\ 0 & 1 \end{pmatrix}.$$

However, we can put necessary and sufficient conditions on the field and quaternion algebra to guarantee that we do get a discrete subgroup, indeed a lattice, in $SL(2, \mathbb{R})$. This is a special case of a much more general theorem on discrete arithmetic groups [3], [1, Chap. 2], [19, Chap. 4].

Theorem 4.2 (Borel-Harishchandra) *Let k be a real number field. Let $\left(\dfrac{a, b}{k}\right)$ be a quaternion algebra over k such that $\left(\dfrac{a, b}{\mathbb{R}}\right) \cong M_2(\mathbb{R})$. Let $\mathcal{O}$ be an order in $\left(\dfrac{a, b}{k}\right)$. Then $\mathcal{O}^1$ is a lattice in $\left(\dfrac{a, b}{\mathbb{R}}\right)^1 (\cong SL(2, \mathbb{R}))$ if and only if*

1. k is totally real

2. $\forall \, \sigma : k \to \mathbb{R}$, σ *a Galois monomorphism* $\neq Id$, *then* $\left(\dfrac{\sigma(a), \sigma(b)}{\mathbb{R}} \right) \cong \mathcal{H}$.

Let me first describe the terminology in this theorem. Any number field is a simple extension of $\mathbb{Q}$ so that we can find α such that $k = \mathbb{Q}(\alpha)$.

Definition 4.3 $k = \mathbb{Q}(\alpha)$ *is totally real if every root of the minimum polynomial of α is real. In that case, if α' is another root of the minimum polynomial, $\alpha \mapsto \alpha'$ induces a Galois monomorphism $\sigma : \mathbb{Q}(\alpha) \to \mathbb{Q}(\alpha') \subset \mathbb{R}$ and there will be exactly $[k : \mathbb{Q}]$ such monomorphisms.*

Examples

1. $k = \mathbb{Q}(\sqrt{d})$ where d is a positive square-free integer, with one non-trivial monomorphism σ where $\sigma(a + b\sqrt{d}) = a - b\sqrt{d}$.

2. $k = \mathbb{Q}(\cos 2\pi/7)$, with three monomorphisms, $\sigma_1 = Id$, and σ_2, σ_3 induced by $\sigma_2(\cos 2\pi/7) = \cos 4\pi/7$, $\sigma_3(\cos 2\pi/7) = \cos 6\pi/7$.

Note that $k = \mathbb{Q}(2^{1/3})$ is not totally real as the roots of the minimum polynomial are $2^{1/3}, e^{2\pi i/3}2^{1/3}, e^{4\pi i/3}2^{1/3}$.

Why then should these number theoretic conditions in the above theorem imply the topological condition of discreteness? Using a particular example, I will sketch the argument and comment on the general case.

Let $k = \mathbb{Q}(\sqrt{2})$ and $A = \left(\dfrac{-1, \sqrt{2}}{\mathbb{Q}(\sqrt{2})} \right)$. Note that $\left(\dfrac{-1, \sqrt{2}}{\mathbb{R}} \right) \cong M_2(\mathbb{R})$. Let $\mathcal{O} = \mathbb{Z}[\sqrt{2}][1, i, j, ij]$ so that $\mathcal{O}$ is an order in A. Then k is totally real and $\sigma(a + b\sqrt{2}) = a - b\sqrt{2}$. Thus $\left(\dfrac{\sigma(-1), \sigma(\sqrt{2})}{\mathbb{R}} \right) = \left(\dfrac{-1, -\sqrt{2}}{\mathbb{R}} \right) \cong \mathcal{H}$. Thus the conditions of the theorem are satisfied and we want to show that $\mathcal{O}^1$ is discrete in $\left(\dfrac{-1, \sqrt{2}}{\mathbb{R}} \right)^1$.

Sketch proof: Suppose not. Then $\exists$ a sequence $\alpha_n \in \mathcal{O}^1$ such that $\alpha_n \to 1$. Let $\alpha_n = x_0^{(n)} + x_1^{(n)}i + x_2^{(n)}j + x_3^{(n)}ij$ where $x_i^{(n)} \in \mathbb{Z}[\sqrt{2}]$. Now consider the sequence $\sigma(\alpha_n) = \sigma(x_0^{(n)}) + \sigma(x_1^{(n)})i' + \sigma(x_2^{(n)})j' + \sigma(x_3^{(n)})i'j'$ where $\{1, i', j', i'j'\}$ is the basis of $\left(\dfrac{-1, -\sqrt{2}}{\mathbb{R}} \right)$. Now $\sigma(\alpha_n) \in \left(\dfrac{-1, -\sqrt{2}}{\mathbb{R}} \right)^1 \cong \mathcal{H}^1 \cong S^3$ which is *compact*. So, for all n, $\sigma(x_0^{(n)})$, $\sigma(x_1^{(n)})$, $\sigma(x_2^{(n)})$, $\sigma(x_3^{(n)})$ are bounded. By assumption, the elements $x_0^{(n)}$, $x_1^{(n)}$, $x_2^{(n)}$, $x_3^{(n)}$ are also bounded for all large n. Suppose e.g. $x_0^{(n)} = a_n + b_n\sqrt{2}$, $a_n, b_n \in \mathbb{Z}$. Thus for infinitely many values of n, both $a_n + b_n\sqrt{2}$ and $a_n - b_n\sqrt{2}$ are bounded. So a_n and b_n are bounded for infinitely many values of n. Contradiction. $\square$

The critical point of this proof is that there is an infinite sequence $\{x_n\}$ of algebraic integers all of whose conjugates are bounded - this is implied by the compactness condition as described in this proof which is consequent on condition 2. of the Borel-Harishchandra Theorem. If the x_n are themselves bounded, then the symmetric polynomials in all the roots of the minimum polynomial of x_n are bounded i.e. all the (integral) coefficients of the minimum polynomial are bounded. Finally the degrees of these minimum polynomials are bounded by $[k : \mathbb{Q}]$. There are clearly only finitely many such polynomials and so the contradiction arrived at in the sketch proof extends to the general case.

Thus conditions 1. and 2. of Theorem 4.2 show that $\mathcal{O}^1$ is discrete in $\left(\dfrac{a,b}{k}\right)^1 \cong SL(2,\mathbb{R})$. It is a much deeper result to show that the image of $\mathcal{O}^1$ has finite covolume in $SL(2,\mathbb{R})$.

Returning to the specific illustrative example used in the sketch proof, it is straightforward to describe the image of $\mathcal{O}^1$. For this we need an isomorphism $\left(\dfrac{-1,\sqrt{2}}{\mathbb{R}}\right) \to M_2(\mathbb{R})$. This is given by

$$1, i, j, ij \to \begin{pmatrix} 1 & 0 \\ 0 & 1 \end{pmatrix}, \quad \begin{pmatrix} 0 & 1 \\ -1 & 0 \end{pmatrix}, \quad \begin{pmatrix} 2^{1/4} & 0 \\ 0 & -2^{1/4} \end{pmatrix}, \quad \begin{pmatrix} 0 & -2^{1/4} \\ -2^{1/4} & 0 \end{pmatrix}.$$

Under this mapping the order $\mathcal{O}$ becomes

$$\left\{ \begin{pmatrix} x_0 + x_2 2^{1/4} & x_1 - x_3 2^{1/4} \\ -x_1 - x_3 2^{1/4} & x_0 - x_2 2^{1/4} \end{pmatrix} \mid x_i \in \mathbb{Z}[\sqrt{2}] \right\}. \tag{2}$$

See also related representations in [11, Chap. 3].

All this finally leads to the definition of an arithmetic Fuchsian group:

Definition 4.4 *Let* $A = \left(\dfrac{a,b}{k}\right)$ *be a quaternion algebra over a totally real number field* k, *such that there is an isomorphism* $\rho : \left(\dfrac{a,b}{\mathbb{R}}\right) \to M_2(\mathbb{R})$ *and such that* $\left(\dfrac{\sigma(a),\sigma(b)}{\mathbb{R}}\right) \cong \mathcal{H}$ *for every Galois monomorphism* $\sigma : k \to \mathbb{R}, \sigma \neq Id$. *Let* $\mathcal{O}$ *be an order in* A. *Then any subgroup of* $SL(2,\mathbb{R})$ *(or* $PSL(2,\mathbb{R})$*) which is commensurable with some such* $\rho(\mathcal{O}^1)$ *is an arithmetic Fuchsian group.*

Thus arithmetic Fuchsian groups are lattices by the Borel-Harishchandra Theorem, so the definition does not depend on the choice of order in A, nor does it depend on the choice of representation ρ. Arithmetic Fuchsian groups can also be defined in terms of quadratic forms over totally real number fields [3, 7]. The class of groups so defined coincides with the class just described, with the initial link being provided by the Adjoint map $Ad : \left(\dfrac{a,b}{k}\right)^1 \to SO(A_0, n)$

where $SO(A_0, n)$ is the special orthogonal group of the three dimensional quadratic subspace $A_0 = \langle i, j, ij \rangle$ of A of pure quaternions where the quadratic form is the restriction of the reduced norm n [9, Chap. 3].

5   Consequences

In general, finitely generated Fuchsian groups of finite covolume are well-understood in that there is a structure theorem for their group presentations, a standard form for their fundamental regions, their volume and torsion are easily determined and their totality describable via Teichmüller theory. So one might ask oneself - why consider this rather specialised number theoretically defined class of groups whose geometric invariants are not at all transparent. In addition, this is rather a small class of groups in that there are only finitely many conjugacy classes of arithmetic Fuchsian groups of any given signature [18]. Indeed, for any $K > 0$, there are only finitely many conjugacy classes of arithmetic Fuchsian groups whose covolumes are less than K [2].

In defence of arithmetic Fuchsian groups, their geometric invariants can, in essence, be deduced directly from the arithmetic data which goes in to their definition. Furthermore, instead of being described by a set of generators, the totality of elements in an arithmetic Fuchsian group is known as we have just shown by the illustrative example at (2) in the last section. Thus, in particular, these arithmetic Fuchsian groups can show the existence of Fuchsian groups with prescribed properties. This is illustrated in [20] where the existence of isospectral non-isometric compact hyperbolic surfaces is given by arithmetic examples. The existence of these arithmetic Fuchsian groups is governed by a strong existence and classification theorem for quaternion algebras over number fields. These points will now be briefly commented on with illustrative theorems and examples.

First consider the classification theorem for quaternion algebras. Recall that, for quaternion algebras over the reals, there are only two possibilities viz. $M_2(\mathbb{R})$ and $\mathcal{H}$. Over any $\mathcal{P}$-adic field $k_\mathcal{P}$ a similar dichotomy holds in that a quaternion algebra over $k_\mathcal{P}$ is either $M_2(k_\mathcal{P})$ or a unique quaternion division algebra [19, Chap. 2], [12, Chap. 17], [9, Chap. 6]. For any totally real number field k, quaternion algebra A over k and valuation v on k (real or $\mathcal{P}$-adic), then A is said to be *ramified* at v if the quaternion algebra $A \otimes_k k_v$ over k_v ($\cong \mathbb{R}$ or $k_\mathcal{P}$) is a division algebra. Thus in this terminology, the condition on the quaternion algebra A to give rise to an arithmetic Fuchsian group as described in definition 4.4 is that A should be ramified at all real places except one, corresponding to the identity. This follows as every real valuation v on k is of the form $v(\alpha) = |\sigma(\alpha)|$, $\alpha \in k$, $\sigma : k \to \mathbb{R}$ is a Galois monomorphism and $|.|$ the usual absolute value on $\mathbb{R}$. Local-global arguments then give the following classification theorem for quaternion algebras over number fields [19, Chap. 3], [12, Chap. 18]:

Theorem 5.1 *Let k be a totally real number field and A a quaternion algebra over k. Let $\Delta(A)$ denote the set of places (equivalence classes of valuations) at which A is ramified. Then*

- *$\Delta(A)$ is finite of even cardinality.*

- *For any finite set S of places on k with even cardinality, there exists a quaternion algebra A over k with $\Delta(A) = S$.*

- *If A_1, A_2 are quaternion algebras over k, then the quaternion algebras A_1, A_2 are isomorphic over k if and only if the sets $\Delta(A_1), \Delta(A_2)$ are equal.*

Now let us consider how the arithmetic data can be used to determine geometrical information. First of all, topological information is also carried by the nature of the quaternion algebra

Theorem 5.2 *Let the quaternion algebra A over the field k give rise to the arithmetic Fuchsian group Γ. Then Γ is non-cocompact if and only if $k = \mathbb{Q}$ and $A \cong M_2(\mathbb{Q})$. In this case, Γ is commensurable in the wide sense with the classical modular group.*

Thus the classical example with which we started this discussion is exceptional in that it corresponds to the single wide commensurability class of non-cocompact arithmetic Fuchsian groups. By the Classification theorem 5.1, there are infinitely many isomorphism classes of quaternion algebras (even over $\mathbb{Q}$) which give rise to arithmetic Fuchsian groups and different isomorphism classes of algebras give rise to distinct wide commensurability classes of Fuchsian groups [17].

In any given quaternion algebra over a number field, the orders are partially ordered by inclusion and each quaternion algebra has finitely many conjugacy classes of *maximal orders* $\mathcal{O}$ [13, Chap. 6], [19, Chap. 4]. The covolumes of the groups $P\rho(\mathcal{O}^1)$, for any maximal order in A, are independent of the choice of maximal order and can be read off from the arithmetic data defining the quaternion algebra [15, 2], [19, Chap. 5].

Theorem 5.3 *Let A be a quaternion algebra over the totally real number field k such that A is ramified at all real places except one. Let ρ be a k-representation of A into $M_2(\mathbb{R})$, and let $\mathcal{O}$ be a maximal order on A. Then the hyperbolic covolume of $P\rho(\mathcal{O}^1)$ is given by the formula*

$$Vol(P\rho(\mathcal{O}^1)) = \frac{8\pi \zeta_k(2) \Delta_k^{3/2} \prod_{\mathcal{P} \in \Delta(A)_f}(N(\mathcal{P}) - 1)}{(4\pi^2)^{[k:\mathbb{Q}]}}. \tag{3}$$

In this formula $\zeta_k(2)$ is the Dedekind zeta function evaluated at 2 and Δ_k is the (absolute) discriminant of the field k. Also $\Delta(A)_f$ is the set of prime ideals $\mathcal{P}$ in k such that A is ramified at the valuation on k corresponding to $\mathcal{P}$. Finally, $N(\mathcal{P})$, the norm of $\mathcal{P}$ is the cardinality of the finite field $R_k/\mathcal{P}$.

Examples

1. $\mathcal{O} = M_2(\mathbb{Z})$ is a maximal order in $A = M_2(\mathbb{Q})$ for which $\Delta(A)_f = \emptyset$.
   Thus (3) yields $Vol(PSL(2,\mathbb{Z})) = \dfrac{8\pi\pi^2/6}{4\pi^2} = \dfrac{\pi}{3}$.

2. In the illustrative example used earlier, $k = \mathbb{Q}(\sqrt{2})$ and $A = \left(\dfrac{-1,\sqrt{2}}{\mathbb{Q}}\right)$
   is ramified at the non-identity real place. By the parity requirement of Theorem 5.1, it must also be ramified at some $\mathcal{P}$-adic place. By standard results on quadratic forms e.g. [9, Chap. 3] it follows that A is also ramified at the prime ideal $\sqrt{2}R_k$ which has norm 2. Thus for a maximal order $\mathcal{O}$ in A, the cocompact group $P\rho(\mathcal{O}^1)$ will have covolume $\dfrac{8\pi\zeta_{\mathbb{Q}(\sqrt{2})}8^{3/2}}{(4\pi^2)^2} = \dfrac{\pi}{6}$ [17]. It should be noted that the order which was described earlier, $\mathbb{Z}[\sqrt{2}][1, i, j, ij]$, is not a maximal order.

Although the group $PSL(2,\mathbb{Z})$ is a maximal discrete Fuchsian group, it is not true in general that the groups $P\rho(\mathcal{O}^1)$ are maximal discrete Fuchsian groups. The covolumes of certain maximal arithmetic Fuchsian groups are given in [17]. A complete analysis of the distribution of arithmetic Fuchsian groups and their covolumes in a commensurability class is made in [2]. This contains a formula for the smallest covolume of an arithmetic Fuchsian group within a commensurability class and a description of the infinitely many conjugacy classes of maximal discrete groups in such a commensurability class. This last situation cannot arise for non-arithmetic groups by Margulis theorem. For this, the commensurator of a subgroup Γ of $PSL(2,\mathbb{R})$ is defined to be

$$Comm(\Gamma) = \{x \in PSL(2,\mathbb{R}) \mid \Gamma \text{ and } x\Gamma x^{-1} \text{ are commensurable}\}.$$

Theorem 5.4 (Margulis) *Let Γ be a lattice in $PSL(2,\mathbb{R})$. Then Γ is non-arithmetic if and only if $[Comm(\Gamma) : \Gamma] < \infty$.*

Again this is a special case of a much more general result. For a proof see [21, Chap. 6].

Theorem 5.3 above gives critical information on the signature of the group $P\rho(\mathcal{O}^1)$ when $\mathcal{O}$ is a maximal order. To determine the complete signature, it is necessary to determine the torsion in $P\rho(\mathcal{O}^1)$. A primitive element of order n will have trace $2\cos\pi/n$ and so correspond to an element $x \in \mathcal{O}^1$ which satisfies

$$x^2 - (2\cos\pi/n)x + 1 = 0. \tag{4}$$

In particular, of course, k must contain $\mathbb{Q}(\cos \pi/n)$ but more especially, an order in the field $k(x)$ must embed in the order $\mathcal{O}$. The number of conjugacy classes of elements of order n in $P\rho(\mathcal{O}^1)$ can then be obtained by counting the number of embeddings of orders in rings of integers in extensions of the field k into the maximal order $\mathcal{O}$ in A. In general, this is rather complicated [19, Chap. 4],[4] but for many cases the formula below, still involving considerable computation, applies [14]

Theorem 5.5 *Let $\mathcal{O}$ be a maximal order such that $P\rho(\mathcal{O}^1)$ contains an element γ with $\mathrm{tr}(\gamma) = s$ where $\epsilon = (s + \sqrt{s^2 - 4})/2$ is a root of unity. Assume that $\{1, \epsilon\}$ is a relative integral basis for $k(\epsilon)$ over k. Then the number of conjugacy classes of maximal cyclic subgroups generated by elements γ with $\mathrm{tr}(\gamma) = s$ and $n(\gamma) = 1$ is*

$$\frac{h(k(\epsilon))}{h[R^*_{k(\epsilon)} : R^{*\,(2)}_{k(\epsilon)}]} \prod_{\mathcal{P} \in \Delta(A)_f} \left(1 - \left(\frac{k(\epsilon)}{\mathcal{P}}\right)\right)$$

where $R^{*\,(2)}_{k(\epsilon)} = \{\alpha \in R^*_{k(\epsilon)} \mid N_{k(\epsilon)|k}(\alpha) \in R^{*2}_k\}$.

In this formula, h is the class number of k, $h(k(\epsilon))$ the class number of $k(\epsilon)$ and $\left(\dfrac{k(\epsilon)}{\mathcal{P}}\right) = 0, 1, -1$ according as $\mathcal{P}$ is ramified, splits or is inert in the extension $k(\epsilon) \mid k$.

Example Consider again the classical example of $PSL(2, \mathbb{Z})$. For elements of order 2, $\epsilon = i$, $h = h(\mathbb{Q}(i)) = 1$, $R^*_{\mathbb{Q}(i)} = R^{*\,(2)}_{\mathbb{Q}(i)}$ and $\Delta(A)_f = \emptyset$. Thus there is one class of order 2 and similiarily one of order 3. Note that combining this with the results of Theorems 5.2, 5.3 shows that the arithmetic data yields the complete signature.

This type of result is not restricted to the cases of elements of finite order so that in (4), $2 \cos \pi/n$ can be replaced by the traces of hyperbolic elements. Thus these counting arguments can also be used to obtain information on geodesics and spectra in these arithmetic Fuchsian groups e.g [20].

Finally, if you want to take advantage of these results, you would want to know whether or not your favourite Fuchsian group is arithmetic. A characterisation of arithmetic Fuchsian groups in terms of their elements enables this to be done. Thus suppose $\Gamma \subset SL(2, \mathbb{R})$ is a lattice and define $\Gamma^{(2)} = \langle \gamma^2 : \gamma \in \Gamma \rangle$. Let $k\Gamma = \mathbb{Q}(\mathrm{tr}\Gamma^{(2)})$ and $A\Gamma = \{\sum a_i \gamma_i : \gamma_i \in \Gamma^{(2)}, a_i \in k\Gamma\}$. Then $A\Gamma$ is a quaternion algebra over $k\Gamma$.

Theorem 5.6 *With notation as above, Γ is arithmetic if and only if*

- *$k\Gamma$ is a totally real number field,*

- *the set $\mathrm{tr}\Gamma$ consists of algebraic integers,*

- *$A\Gamma$ is ramified at all real places except one.*

This result is a version of a characterisation theorem proved in [16] and later subjected to a number of variations [10, 8, 6] which make its application straightforward.

References

[1] A. Borel, *Introduction aux Groupes Arithmètiques*, Hermann, Paris, 1969.

[2] A. Borel, *Commensurability classes and volumes of hyperbolic 3-manifolds*, Ann. Sc. Norm. Sup. Pisa, **8**, (1981), 1 – 33.

[3] A. Borel and Harishchandra, *Arithmetic subgroups of algebraic groups*, Annals of Maths. **75**, (1962), 485–535.

[4] T. Chinburg and E. Friedman, *Torsion in maximal arithmetic subgroups of PGL(2, $\mathbb{C}$)*, Preprint.

[5] P.M. Cohn, *Algebra*, Vol. 3, Wiley, Chichester, 1993.

[6] F.W. Gehring, C. Maclachlan, G. Martin and A.W. Reid, *Arithmeticity, discreteness and volume*, Trans. Amer. Math. Soc. **349**, (1997), 3611–3643.

[7] H. Hilden, M. Lozano and J Montesinos-Amilibia, *On the Borromean orbifolds: Geometry and Arithmetic*, Topology '90, Ed B. Apansov et al, De Gruyter, Berlin, 1992, 133–167.

[8] H. Hilden, M. Lozano and J. Montesinos-Amilibia, *A characterization of arithmetic subgroups of SL(2, $\mathbb{R}$) and SL(2, $\mathbb{C}$)*, Math. Nachrichten, **159**, 1992, 245–270.

[9] T.Y. Lam, *Algebraic theory of quadratic forms*, Benjamin, Reading, Mass., 1973.

[10] C. Maclachlan and G. Rosenberger, *Two-generator arithmetic Fuchsian groups*, Math. Proc. Camb. Phil. Soc. **93**, (1983), 383–391.

[11] W. Magnus, *Non-Euclidean tesselations and their groups*, Academic Press, New York, 1974.

[12] R.S Pierce, *Associative Algebras*, Grad. Texts in Maths, Vol. 88, Springer-Verlag, New York, 1982.

[13] I. Reiner, *Maximal Orders*, Academic Press, London, 1975.

[14] V. Schneider, *Die elliptischen Fixpunkte zu Modulgruppen in Quaternionenschiefskörpern*, Math. Ann. **217**, (1975), 29–45.

[15] H. Shimizu, *On zeta functions of quaternion algebras*, Annals of Maths **81**,(1965), 166-193.

[16] K. Takeuchi, *A characterization of arithmetic Fuchsian groups*, J. Math. Soc. Japan **27**, (1975), 600–612.

[17] K. Takeuchi, *Commensurability classes of arithmetic triangle groups*, J. Fac. Sc. Univ. Toyko 1A **24**, (1977), 201–212.

[18] K. Takeuchi, *Arithmetic Fuchsian groups with signature* $(1; e)$, J. Math. Soc. Japan **35**, (1983), 381–404.

[19] M-F Vigneras, *Arithmétiques des Algèbres de Quaternions*, Lect. Notes in Maths 800, Springer-Verlag, Berlin, 1980.

[20] M-F Vigneras, *Variétés riemannienes isospectrales et non isométriques*, Ann. Maths. **112**, (1980), 21–32.

[21] R.J. Zimmer, *Ergodic theory and semisimple groups*, Birkhäuser, Boston, 1984.

Department of Mathematical Sciences
King's College
University of Aberdeen
Aberdeen AB24 3UE
Scotland.

RIEMANN SURFACES, BELYI FUNCTIONS AND HYPERMAPS

David Singerman

Introduction. Belyi's Theorem and the associated theory of dessins d'enfants has recently played an important rôle in Galois theory, combinatorics and Riemann surfaces. In this mainly expository article we describe some consequences for Riemann surface theory. It is organised as follows: in §1 we describe the ideas of critical points and critical values which leads in §2 to the definition of a Belyi function. In §3 we state Belyi's Theorem and define a Belyi surface. In §4 the close connection with triangle groups is described and this leads in §5 to an account of maps and hypermaps (or dessins d'enfants) on a surface. These are closely related to Belyi surfaces but in order to investigate this connection we introduce the idea of a smooth Belyi surface and a platonic surface in §6. The only new result in this article is Theorem 7.1 which describes the connection between regular maps and thir underlying Riemann sufaces.

1 Preliminaries

The definition of a Riemann surface is given in Beardon's lecture in this volume and the important ideas of uniformization are also discussed there. In the definition of a Riemann surface the *transition functions* $\phi_i \phi_j^{-1}$ are complex analytic and this allows us to describe most of the important ideas of complex analysis on a Riemann surface. The simplest compact Riemann surface is the *Riemann sphere*. Topologically, this is just the 2-sphere and we make it into a Riemann surface by placing it in $\mathbf{R}^3$ so that the equator of the sphere is the unit circle in the complex plane $\mathbf{C}$. We choose a covering of the sphere by two open sets N_1 and N_2 where N_1 is the sphere without the north pole $(0, 0, 1)$ and N_2 is the sphere without the south pole $(0, 0, -1)$. The complex coordinates in N_1 are given by stereographic projection from the north pole; in other words, if we draw a straight line from the north pole to a point $P \in N_1$ which cuts the complex plane at z then we give P the complex coordinate z. If $P \in N_2$ as well and we stereographically project from the south pole then we find that the coordinate is $1/z$. As $z \mapsto 1/z$ is complex analytic on $\mathbf{C} - \{0\}$ we have made the sphere into a Riemann surface which we denote by Σ and by the Uniformization Theorem, as given in Beardon's lecture, every Riemann surface of genus zero (that is, homeomorphic to the sphere) is conformally equivalent to Σ. Note that the Riemann sphere is a model of $\mathbf{C} \cup \{\infty\}$, with the north pole representing ∞.

1.1 Meromorphic functions on a Riemann surface.

A *meromorphic function* on the Riemann surface $\mathcal{R}$ is a holomorphic function $\chi : \mathcal{R} \to \Sigma$. Note that in some works analytic is used instead of holomorphic.

In this article we will pay particular attention to compact Riemann surfaces. In this case, the meromorphic functions are easy to visualise because of the following result. ([FK§1.1.6).

Proposition 1.1 *Every non-constant meromorphic function χ on a compact Riemann surface $\mathcal{R}$ takes each value the same number of times counting multiplicity.*

We can explain the term 'multiplicity' with a simple example. If $\mathcal{R} = \Sigma$ then the function $z \mapsto z^n$ takes every value except for 0 and ∞ precisely n times. However $z^n = 0$ has an n fold root at $z = 0$ and at ∞ where we have to use the open set N_2 and the homeomorphism $z \mapsto 1/z$ as chart, we see that $z^n = \infty$ has an n-fold root at ∞. Thus $z \mapsto z^n$ takes every value n times counting multiplicity. Now using Laurent expansions we can show that any meromorphic function $\chi : \mathcal{R} \to \Sigma$ is locally like $z \mapsto z^n$ in some system of local coordinates so this simple example can be extended to give a proof of the proposition.

A non-constant meromorphic function χ which takes every value n times, counting multiplicity, is called a meromorphic function of *degree n*. Thus for most points $z \in \Sigma$, the set $\chi^{-1}(z)$ has n distinct points.

Definition. If χ is a meromomorphic function of degree n and if $\chi^{-1}(z_0)$ has less than n distinct points then z_0 is called a *critical value* of χ. We also call any point in $\chi^{-1}(z_0)$ a *critical point* of χ.

Example 1.2 *The points 0 and ∞ are the critical values of the function $z \mapsto z^n$, (as well as being critical points in this case.)*

A meromorphic function ϕ on the sphere is a rational function ([JS2], Theorem 1.4.1) and in this case it is easy to find the critical points and values. If $z_0 \in \Sigma$, and both $z_0 \neq \infty$ and $\phi(z_0) \neq \infty$ then $\phi(z_0) = w_0$ with multiplicity n if the first $(n-1)$ derivatives of ϕ vanish at z_0 but $\phi^{(n)}(z_0) \neq 0$. In particular, in these cases, z_0 is a critical point of ϕ if $\phi'(z_0) = 0$ and $\phi(z_0)$ is the corresponding critical value.

Example 1.3 *Let m and n be integers with m, n, $m + n \neq 0$. We will investigate the critical points and values of*

$$\pi_{m,n}(z) = z^m(1 - z)^n.$$

First of all, $\pi'_{m,n}(z) = z^{m-1}(1 - z)^{n-1}(m - (m + n)z)$ and so $m/(m + n)$ is a critical point with corresponding critical value

$$\mu = \frac{m^m n^n}{(m + n)^{m+n}}.$$

The only possible other finite critical points are 0 and 1 and in these cases the possible critical values are 0 and ∞. Thus $\pi_{m,n}(z)$ has critical points lying in the set $\{0, 1, \frac{m}{(m+n)}, \infty\}$, and critical values lying in the set $\{0, \mu, \infty\}$.

2 Belyi functions

Definition. A meromorphic function $\beta : \mathcal{R} \to \Sigma$ is called a *Belyi function* if the critical values of β lie in the set $\{0, 1, \infty\}$.

Example 2.1 *If $\mathcal{R} = \Sigma$, then $z \mapsto z^n$ is a Belyi function.*

Example 2.2 *If $\mathcal{R} = \Sigma$, then*

$$\beta_{m,n} : z \mapsto \frac{1}{\mu}z^m(1 - z)^n$$

is a Belyi function as follows from example 1.3.

There are many more interesting examples of Belyi functions on the sphere. For example, if $T_n(z)$ is the nth Chebyshev polynomial (which is defined by $T_n(z) = \cos(n \cos^{-1} z)$) then $1 - T_n(z)^2$ is a Belyi function, and this example has been extended to form generalized Chebyshev or *Shabat* polynomials which have an interesting relationship to plane trees ([SZ]). However, our main interest in this article is with Riemann surfaces and as there is only one complex structure on the sphere we shall be more concerned with examples of Belyi functions on Riemann surfaces of genus ≥ 1.

The next example comes from the theory of algebraic curves. We consider the Fermat curve

$$x^n + y^n = 1.$$

We build the Riemann surface $\mathcal{R}$ of this curve in the usual way by considering the sphere Σ as the x-sphere and the points of the surface $\mathcal{R}$ as having "coordinates" (x, y). There is a projection $\pi : \mathcal{R} \to \Sigma$ defined by $\pi(x, y) = x$. Now given $x_0 \in \Sigma$, we can find its inverse images by noting that $\pi(x_0, y) = x_0$, if $x_0^n + y^n = 1$. Thus there are n inverse images, *unless* $x_0^n = 1$, or $x_0 = \infty$ and so there are $(n + 1)$ critical values of π and π is not a Belyi function if $n > 2$. (However, we can now use the Riemann-Hurwitz formula to show that $\mathcal{R}$ has genus $(n - 1)(n - 2)/2$).

Example 2.3 *Consider the projection $\beta(x, y) = x^n$. For all values $u = x^n \in \Sigma \setminus \{0.1.\infty\}$ there are n values of x and then for each value of x there are n values of y, so that β is an n^2-sheeted cover. For which values of u are there less than n^2 images? Clearly this is the case for $u = 0$ and $u = 1$. These points have only n inverse images. To deal with ∞ it is more convenient to pass to projective coordinates by putting $x = \frac{X}{Z}, y = \frac{Y}{Z}$. Now $X^n + Y^n = Z^n$, and $(X, Y, Z) \equiv (\lambda X, \lambda Y, \lambda Z)$ for all non-zero scalars λ. The point at ∞ corresponds to $Z = 0$ and then $(Y/X)^n = -1$, giving n values of Y/X so that ∞ also has n inverse images and so is a critical value. Thus β is a Belyi function.*

In this next example we show how to construct Belyi functions from triangle groups. First, we recall the definition of a triangle group. Let l, m, n be integers with $\frac{1}{l} + \frac{1}{m} + \frac{1}{n} < 1$. Let T denote the triangle in the hyperbolic plane with angles $\pi/l, \pi/m$, and π/n, and Δ^* denote the group generated by reflections in the sides of T. Then the subgroup of index 2 in Δ^* consisting of conformal transformations is a Fuchsian group known as a *triangle group* and denoted by $\Delta = \Delta(l, m, n)$. It is known and easy to show that this triangle group is determined up to conjugacy in the group of all conformal homeomorphisms of $\mathbf{H}$ (where $\mathbf{H}$ is the upper half-plane) by the integers l, m, n. We note that $\Delta(l, m, n)$ is generated by three elliptic generators x_1, x_2, x_3 each with a unique fixed point in $\mathbf{H}$. A presentation of $\Delta(l, m, n)$ is

$$\langle x_1, x_2, x_3 | x_1^l = x_2^m = x_3^n = x_1 x_2 x_3 = 1 \rangle \tag{2.1}$$

Now let Γ be a subgroup of finite index in Δ. Then Γ is a Fuchsian group and then $\mathbf{H}/\Gamma$ is a compact Riemann surface. (This was mentioned in Beardon's lecture in the case when Γ has no elliptic elements, but the result is true even if Γ has elliptic elements, see Theorem 5.9.1 of [JS2]). Now the quotient space $\mathbf{H}/\Delta$ is a sphere and the natural projection $\pi : \mathbf{H}/\Gamma \mapsto \mathbf{H}/\Delta$ has at most 3 critical values; these occur at the projections of the fixed points of the elliptic generators of Δ, but possibly if this fixed point is a fixed point of an elliptic generator of Γ, then there will be less than 3 critical values. However we can now apply a Möbius transformation σ mapping the critical values into $\{0, 1, \infty\}$ and then

Example 2.4 $\beta := \sigma\pi$ *is a Belyi function from* $\mathbf{H}/\Gamma$ *to* $\mathbf{H}/\Delta$.

Note: In the above description we have only considered triangle groups in the hyperbolic plane. However, all the above applies to triangle groups such as $\Delta(2, 4, 4)$ in the Euclidean plane or even the finite triangle groups on the sphere such as $\Delta(2, 3, 5)$.

So far we have only considered triangle groups with finite periods l, m, n. However if we consider hyperbolic triangles with a vertex on $\mathbf{R} \cup \{\infty\}$, then at that vertex there is an angle equal to zero and this corresponds to an infinite period. In the presentation (2.1) a relation $x_i^\infty = 1$ is regarded as vacuous, so that triangle groups with an infinite period are free products of cyclic groups, including the important case $\Delta(\infty, \infty, \infty) \cong F_2$, the free group on two generators. Note that an elliptic element of "infinite period" fixes a unique point on $\mathbf{R} \cup \{\infty\}$ and so is a parabolic element. Triangle groups with infinite periods are important because some arithmetic groups are of this form. In particular, the classical modular group $G = PSL(2, \mathbf{Z})$ consists of the Möbius transformations

$$T : z \mapsto \frac{az + b}{cz + d}, \quad a, b, c, d \in \mathbf{Z}, \quad ad - bc = 1.$$

We also consider the subgroups

$$G_0(2) = \{T \in G | \ c \ \equiv \ 0 \ mod \ (2) \ \}$$

and

$$G(2) = \{T \in G | \ b \ \equiv c \ \equiv 0 \ mod(2) \ \}.$$

As triangle groups we have $G \cong \Delta(2, 3, \infty)$, $G_0(2) \cong \Delta(2, \infty, \infty)$, $G(2) \cong \Delta(\infty, \infty, \infty)$.

3 Why are Belyi functions important?

If we consider an irreducible complex algebraic curve $f(w, z) = 0$ completed at ∞ in a suitable way then it defines a compact Riemann surface. This is a classical construction, described in some detail in [JS2]. We have already seen the example of the Fermat curves above. Remarkably, the converse of this result holds; every compact Riemann surface is the Riemann surface of some algebraic curve. For a deeper clarification and proof of these results see [G] or [FK]. Now let K be some subfield of the field $\mathbf{C}$ of complex numbers. Then we say that a compact Riemann surface $\mathcal{R}$ is defined over K if it is isomorphic (that is birationally equivalent) to the Riemann surface of some curve $f(w, z) = 0$ defined over K. (That is the coefficients of the polynomial $f(w, z)$ lie in the field K.)

Theorem 3.1 *(Belyi [B]) A compact Riemann surface $\mathcal{R}$ is defined over the field $\overline{\mathbf{Q}}$ of algebraic numbers if and only if there exists a Belyi function β : $\mathcal{R} \to \Sigma$.*

In [B], a short paper devoted to Galois groups, Belyi proved that if $\mathcal{R}$ is defined over $\overline{\mathbf{Q}}$ then a Belyi function $\beta : \mathcal{R} \to \Sigma$ exists. For the converse he quoted Weil's rigidity theorem of 1956. Recently, Wolfart [Wo] has clarified Weil's proof. Referring to Belyi's result, Grothendieck in his Esquisse d'un programme,([SL]) writes *'never, without a doubt, was such a deep and disconcerting result proved in so few lines!'*

Definition. A Riemann surface $\mathcal{R}$ defined over $\overline{\mathbf{Q}}$ will be called a *Belyi surface.*

Instead of giving Belyi's proof which may be found in [B, JS3, Jo] we illustrate it by concentrating on the special case of the elliptic curves of the form

$$w^2 = z(z - 1)(z - \lambda).$$

If we let $\pi(z, w) = z$, then every value of z has two inverse images except for the 4 critical values $0, 1, \lambda$ and ∞. First suppose that λ is a rational number which we write as $\lambda = m/(m + n)$. Then by example 2.2 we find that $\beta_{m,n} \circ \pi$ has critical values in $\{0, 1, \infty\}$ so that $\beta_{m,n} \circ \pi$ is a Belyi function.

Now take $\lambda = 1 + \sqrt{2}$. Then we apply the minimal polynomial of λ, namely $p(z) = -z^2 + 2z + 1$ to take λ to 0, and consider the composition $p \circ \pi$. This is a composition of two functions of degree 2 and so has degree $2 \times 2 = 4$. Note that $p'(z)$ vanishes only for $z = 1$, and thus $p(1) = 2$ is a critical value of p and hence also of $p \circ \pi$. The critical values of $p \circ \pi$ still include $\{0, 1, \infty\}$, but we have introduced a new critical value namely 2. However $2 = 2/(2\text{-}1)$ and then we obtain a Belyi function

$$\beta = \beta_{2,-1} \circ p \circ \pi(z, w) = \frac{(z^2 - 2z - 1)^2}{4z(2 - z)}$$

of degree 8. Belyi's proof amounts to using minimal polynomials to force the critical values into $\mathbf{Q} \cup \{\infty\}$ and then using functions of the form $\beta_{m,n}$ and Möbius transformations to force the critical values into $\{0, 1, \infty\}$.

4   Belyi functions and triangle groups.

We saw in Example 2.4 that if $\Delta = \Delta(l, m, n)$ is a Fuchsian triangle group and if $\Gamma < \Delta$ with finite index then the natural projection $\beta : \mathbf{H}/\Gamma \to \mathbf{H}/\Delta$ is a Belyi function. Conversely, in [CIW] it is proved that all Belyi functions are essentially of this form. In the case where $\frac{1}{l} + \frac{1}{m} + \frac{1}{n} = 1$ we have to replace $\mathbf{H}$ by $\mathbf{C}$ and when $\frac{1}{l} + \frac{1}{m} + \frac{1}{n} > 1$ we have to replace $\mathbf{H}$ by Σ. We let $\mathcal{U}$ denote any one of the spaces $\mathbf{H}$, $\mathbf{C}$ or Σ; these are the 3 simply-connected Riemann surfaces by the Uniformization Theorem. Then Belyi surfaces are precisely the surfaces of the form $\mathcal{U}/\Gamma$ where Γ is a subgroup of finite index in a cocompact triangle group. Now if l', m', n' are multiples of l, m, n respectively then there is an obvious homomorphism $\theta : \Delta(l', m', n') \to \Delta(l, m, n)$ and if Γ is a subgroup of finite index in $\Delta(l, m, n)$ then $\theta^{-1}(\Gamma)$ is a subgroup of the same index in $\Delta(l', m', n')$ and we can show that the quotient surfaces $\mathcal{U}/\Gamma$ and $\mathcal{U}/\Gamma'$ are conformally equivalent. If we make the convention that k divides ∞ for all positive integers k, then we may extend this result to triangle groups with infinite periods, by replacing $\mathbf{H}$ by $\hat{\mathbf{H}}$, the space obtained by adjoining the parabolic fixed points to $\mathbf{H}$. In particular all Belyi surfaces may be represented in the form $\hat{\mathbf{H}}/\Gamma$ where Γ is a subgroup of finite index in $G(2) \cong \Delta(\infty, \infty, \infty)$ and $\hat{\mathbf{H}} = \mathbf{H} \cup \mathbf{Q} \cup \{\infty\}$, for $\mathbf{Q} \cup \{\infty\}$ is the set of parabolic fixed points of the modular group G and hence also for any finite index subgroup in G.

Theorem 4.1 *The following are equivalent:*

1. $\mathcal{R}$ is a Belyi surface. (i.e. $\mathcal{R}$ is defined over $\overline{\mathbf{Q}}$),

2. There is a Belyi function $\beta : \mathcal{R} \to \Sigma$,

3. $\mathcal{R} \cong \mathcal{U}/\Gamma$, where Γ is a finite index subgroup in a cocompact triangle group,

4. $\mathcal{R} \cong \hat{\mathbf{H}}/\Gamma$, where Γ is a finite index subgroup in $G(2)$,

5. $\mathcal{R} \cong \hat{\mathbf{H}}/\Gamma$, where Γ is a finite index subgroup in G,

6. $\mathcal{R} \cong \hat{\mathbf{H}}/\Gamma$, where Γ is a finite index subgroup in $G_0(2)$.

Proof. The equivalence of 1 and 2 is Belyi's Theorem. The equivalence of 2, 3, and 4 follows from the discussion before the statement of the Theorem. If we assume 4, then as $G(2) < G$, any subgroup of $G(2)$ is a subgroup of G and thus $4 \implies 5$. If we assume 5, then as there is a canonical homomorphism θ from $G(2)$ to G, then if $\mathcal{R} \cong \mathbf{H}/\Gamma$, where Γ is a finite index subgroup in G, then $\mathcal{R} \cong \mathbf{H}/\theta^{-1}(\Gamma)$, where Γ is a finite index subgroup in $G(2)$. Similarly, we can prove the equivalence of the statements 4 and 6.

5   Maps and hypermaps on Riemann surfaces

If we attempt to study all Riemann surfaces of genus $g > 1$, then by the Uniformization Theorem we could study all torsion-free Fuchsian groups that represent a surface of genus g, and we then find ourselves studying Teichmüller space (of dimension $6g - 6$) and the associated moduli space. If we want to study all Belyi surfaces of genus g then by Theorem 4.1, we restrict ourselves to the study of genus g subgroups of triangle groups, which at first sight may seem a simpler task. However, we no longer restrict ourselves to surface groups, so that the same Riemann surface may be represented by non-conjugate subgroups of a particular triangle group. We might expect the theory to be, in some sense, more discrete, which also fits in nicely with the fact that over $\overline{\mathbf{Q}}$ the number of curves of genus g is countable. The discrete theory that we obtain is that of maps or hypermaps on compact orientable surfaces. A current term (due to Grothendieck; see [SL]) that includes both maps and hypermaps is "dessin d'enfants" or just dessin. We begin by describing the easier and more intuitive idea of a map. (The term "map" is used here in its cartographic sense. It is not a function!) Basically, a map is just a decomposition of a surface into polygonal two-cells or *faces*. For example, we could think of a cube as being a decomposition of the sphere into six square faces. Every map has vertices, edges, and faces and the vertices and edges form a graph embedded into the surface. We must first be careful of the types of graphs that we allow. They can have edges with two vertices, one at either end, (i.e. homeomorphs of the closed unit interval $[0, 1]$) but we also need to consider edges with one vertex. These can be *loops* (homeomorphs of a circle) with the vertex as a distinguished point, or homeomorphs of the interval $[0, 1]$ but with a single vertex at the point corresponding to 0; we call these *free edges*. Free edges do not usually occur in traditional treatments of graph theory, but they are necessary for a complete treatment of maps as we shall see. Graphs which can include multiple edges, loops and free edges were called *allowed graphs* in [JS1]. However, we shall just refer to them as graphs.

Definition 5.1 A *map* $\mathcal{M}$ on a compact surface $\mathcal{S}$ is an embedding of a graph $\mathcal{G}$ into $\mathcal{S}$ such that the components of $\mathcal{S} \setminus \mathcal{G}$ (called the faces of $\mathcal{M}$) are simply-connected.

Definition 5.2. If e is an edge of $\mathcal{M}$ which is not a loop and if v is a vertex of e then the ordered pair (v, e) is called a *dart* of $\mathcal{M}$.

This dart is represented by an arrow on the edge e pointing towards the vertex v. If e is a loop, with the single vertex v, then we can define two darts on e, one of which represents an arrow pointing towards v in an anticlockwise direction and the other representing an arrow pointing towards v in a clockwise direction. In this way, every non-free edge carries two darts, while every free edge carries one dart.

Example 5.1

1. The platonic solids give maps on the sphere. For example, the icosahedron has 12 vertices, 20 faces, 30 edges and 60 darts.

2. Klein's map on a Riemann surface of genus 3. This gives a map with 56 vertices, 24 heptagonal faces, 84 edges and 168 darts. (e.g. see [L]).

3. The star maps of genus 0 consist of a single vertex and n free edges emanating from that vertex. These maps have n darts. They also have a single face.

4. Any plane tree gives a genus zero map with a single face.

Just as we obtain maps from graph embeddings, we can also obtain hypermaps by hypergraph embeddings. A hypergraph contains vertices and edges, but now we allow an edge to contain any finite number of vertices, whereas a graph only contains two or one vertex on every edge. As described in [W], for example, we can define a hypergraph purely abstractly as follows. A hypergraph $\mathcal{H}$ consists of a finite non-empty set $V(\mathcal{H})$ of *vertices*, together with a collection of subsets (not necessarily disjoint) $E(\mathcal{H})$, of *edges* each of whose elements is a subset of $V(\mathcal{H})$.

A *bipartite graph* $\mathcal{B}$ is a graph in which the vertex set V of $\mathcal{B}$ can be written as a disjoint union of two non-empty subsets $V = V_1 \cup V_2$ and such that an edge of $\mathcal{B}$ can only join vertices in V_1 with vertices in V_2.

Given a hypergraph $\mathcal{H}$ we can form its associated bipartite graph $\mathcal{B}(\mathcal{H})$ as follows. The vertex set of $\mathcal{B}(\mathcal{H})$ is of the form $V_1 \cup V_2$ where V_1 is the vertex set of $\mathcal{H}$ and V_2 is the edge set of $\mathcal{H}$. We join $v \in V_1$ to $e \in V_2$ if v lies on the edge e in $\mathcal{H}$.

Example 5.2 *One of the simplest hypergraphs contains 3 vertices and 3 edges with every vertex lying on every edge and every edge containing all 3 vertices. The associated bipartite graph is the complete bipartite graph $K_{3,3}$, consisting of two sets V_1, V_2 of 3 vertices with every vertex of V_1 joined to every vertex of V_2 by a unique edge.*

Example 5.3 *The finite projective plane of order 2 contains 7 vertices and 7 lines with 3 lines on each vertex. It is the projective plane defined over a field of order 2 and the vertices are all 3-tuples of zeros and ones except for (0,0,0), the lines being defined in the usual way in projective geometry. The underlying hypergraph is the Fano plane and the associated bipartite graph is the Heawood graph (See fig. 8-6 of [W]).*

One definition of a hypermap is an embedding of a connected hypergraph into a surface such that the complementary regions (i.e. the surface minus the embedded hypergraph) are simply-connected. The images of the vertices, edges and these complementary regions are known as *hypervertices, hyperedges and hyperfaces*, respectively.

Unlike maps it is not obvious how to best draw a hypermap on a surface, so we give some intrinsic definitions of a hypermap. We can regard these descriptions as geometric representations of a hypermap on an orientable surface.

5.1 The Cori representation

(See([C,CS]). Let $\mathcal{X}$ be a compact orientable surface. A hypermap $\mathcal{H}$ on $\mathcal{X}$ is a triple $(\mathcal{X}, S, A)$, where S, A are closed subsets of $\mathcal{X}$ such that:

- (i) $B = S \cap A$ is a non-empty finite set;

- (ii) $S \cup A$ is connected;

- (iii) Each component of S and each component of A is homeomorphic to a closed disc;

- (iv) Each component of $\mathcal{X} \setminus (S \cup A)$ is homeomorphic to an open disc;

The elements of the set B are called *brins*. The components of S are called hypervertices, the components of A are called hyperedges and the components of $\mathcal{X} \setminus (S \cup A)$ are called hyperfaces. It is clear how this corresponds to a hypergraph embedding as described above. We can think of the hyperedges, hypervertices and hyperfaces as topological polygons whose vertices are the brins. As every brin lies on a unique hypervertex and hyperedge, it follows that the underlying graph whose vertices are the brins in the Cori representation is 4-valent. Also, when we traverse a hyperface, we must traverse arcs that lie alternately in ∂S and ∂A respectively, so that the hyperfaces must always have even valency. When we relate this to Belyi functions, this becomes a little unnatural so we now consider;

5.2 The James representation

(See[J]). Consider a trivalent map $\mathcal{M}$ on $\mathcal{X}$. (This means that every vertex of the underlying graph has degree 3). We call $\mathcal{M}$ a hypermap if the faces are labelled $i = 0, 1, \infty$ so that each edge separates faces of different labels. Around every vertex we have a permutation obtained by following the type of a face obtained by going anticlockwise around a small circle with centre at that vertex. This permutation is either $(0,1,\infty)$ or $(0,\infty,1)$. The vertex is called a brin if the permutation is $(0,1,\infty)$.

The hypervertices, hyperedges and hypervertices are now the regions labelled $0,1,$ ∞ respectively. To go from the Cori representation to the James representation we push the hypervertex and neighbouring hyperedge in the Cori representation slightly together, thereby changing the brin in the Cori representation (which is a vertex) to an edge in the James representation one of whose vertices is a brin, as just described. We notice that every hyperedge and hypervertex in the James representation has now double the number of edges that these have in the Cori representation (as the brins have now become edges) so that every hyperedge, hypervertex and hyperface have even numbers of sides; also, whereas the Cori representation gives a 4-valent graph, the James representation gives a 3-valent graph. In Fig.1(a) and Fig.1(b) we draw the hypermap corresponding to the hypergraph of Example 5.2 in its Cori and James representations respectively.

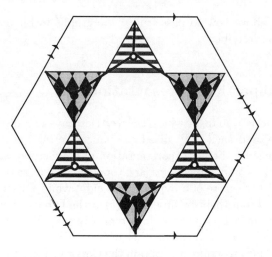

Fig. 1(a)

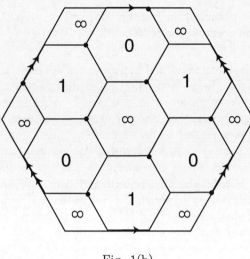

Fig. 1(b)

We explained above how to pass from hypergraphs to bipartite graphs. We use this idea to describe

5.3   The bipartite representation

A *bipartite map* is a map in which the vertices have one of two colours, say black or white, and such that all edges connect vertices of different colours. Suppose we consider the Cori representation of a hypermap as defined above. The hypervertices and hyperedges are topological polygons. If we place a black vertex at the centre of a hypervertex and a white vertex at the centre of a hyperedge and join them by an edge $\alpha(b)$ if the hypervertex and hyperedge intersect at a brin b then we obtain a bipartite map on the surface and the number of edges is equal to the number of brins. (See Fig. 1(a)). Conversely, we can reverse this procedure and obtain the Cori representation of a hypermap out of a bipartite map. Thus we may regard a hypermap as a bipartite map, with the hypervertices being the black vertices, the hyperedges being the white vertices, and the hyperfaces as being the faces of the map.

5.4   The relationship between maps and bipartite maps

In the above paragraph we have discussed ways of passing from a hypermap on a surface to a bipartite map. We shall see that this construction is essentially unique, in that the maps we obtain in these ways from a given hypermap are essentially unique. However, we also note that every map is also a hypermap in which every hyperedge has two sides. (Assuming that every edge has two vertices, this means that in the Cori representation every hyperedge is a digon, and we can identify the edges of this digon to form a single edge.) In the James representation we have to squash down a 4-sided polygon down to a single edge, and in the bipartite representation we just require all white vertices to have valency two, which can then be removed. If we have any edges in the map with just one vertex, the constructions are similar. However, note that if we regard a map as a hypermap then the darts become the brins.

5.5 Hypermaps and permutations

From every hypermap on an orientable surface we can find a pair of permutations. Consider the Cori representation of the hypermap and take a hypervertex V of valency v. We get a v-cycle by following the brins anticlockwise around V. If the hypervertices are $V_1, \ldots, V_h$ of valencies $v_1, \ldots, v_h$ respectively we get a permutation g_0 of the brins as a composition of these disjoint cycles. The order of this permutation l is the least common multiple (lcm) of the lengths of these cycles, namely $[v_1, \ldots, v_h]$. Similarly, if we have hyperedges of lengths $w_1, \ldots, w_e$, we obtain a permutation g_1 of order $m = [w_1, \ldots, w_e]$. If we consider the product permutation $g_0 g_1$ it is easy to see that we go around two edges of a hyperface in a clockwise direction. (Recall that every hyperface in the Cori representation has even valency.) Thus if the hyperfaces have valencies $2y_1, \ldots, 2y_f$, then $g_\infty := g_0 g_1$ has order $n = [y_1, \ldots, y_f]$. Thus if we let G be the group generated by g_0 and g_1, (called the *monodromy group of the hypermap*) then we have, at least, the relations

$$g_0^l = g_1^m = (g_0 g_1)^n = 1.$$

If we put $g_\infty = (g_0 g_1)^{-1}$ then we see that G is generated by g_0, g_1, g_∞ and obeys at least the relations

$$g_0^l = g_1^m = g_\infty^n = g_0 g_1 g_\infty = 1. \tag{5.1}$$

In this case we say that the hypermap has *type* (l, m, n). Note that the connectedness of the hypermap corresponds to G being a transitive group on the brins. As an example, in Fig 1(a) each of the permutations g_0, g_1, g_∞ that we obtain is a product of 3 three-cycles. (See [CS]).

If $m = 2$ then we say that the hypermap is a map, this precisely being the case when every hyperedge has one or two brins. In the following section we shall also see that the converse holds. That is, if we have a finite permutation group G acting transitively on N points, and obeying the relations (5.1) then we can recover a hypermap of type (l, m, n) with N bits.

5.6   Hypermaps and triangle groups

If we compare (2.1) and (5.1) then it is clear that there is an obvious epimorphism $\theta : \Delta(l, m, n) \to G$. Thus if we have a hypermap $\mathcal{H}$ of type (l, m, n) then there is an induced transitive action of the triangle group $\Delta(l, m, n)$ on the brins of $\mathcal{H}$. *Remark on notation. If we permute the periods in $\Delta(l, m, n)$ then we obtain an isomorphic group. However, we often mean by a hypermap of type (l, m, n), that the vertices have valency l, the edges have valency m and the faces have valency n.* If $\mathcal{H}$ has N brins then the stabilizer of a brin b, $\Delta_b(l, m, n)$ has index N in $\Delta(l, m, n)$. We will call this subgroup the *fundamental group of $\mathcal{H}$* and denote it by $\pi_1(\mathcal{H}, b)$. If we change the 'base brin' b then we just pass to a conjugate subgroup in $\Delta(l, m, n)$, and so we can usually omit b and just write $\pi_1(\mathcal{H})$. (In [JS1] we called this subgroup the map subgroup of the hypermap but below we shall give reasons why this new terminology is appropriate and helpful.) First, we need the concept of the *universal hypermap* of type (l, m, n), which lies on one of the 3 simply-connected Riemann surfaces, depending in the usual way on whether $1/l + 1/m + 1/n$ is greater than, equal to or less than 1. It is slightly easier to describe universal maps and this concept can be readily understood by describing the universal map of type (3,2,6). Consider the tessellation of the plane by regular hexagons, 3 meeting at each vertex. Then the edges, vertices and hexagons are just the edges, vertices and faces of the universal map of type (3,2,6). Note that the vertices, edge-centres and face-centres are just the points fixed by the finite subgroups of $\Delta(3, 2, 6)$, and that the stabilizer of a brin (= dart in the case of maps) is trivial. We denote the universal hypermap of type (l, m, n) by $\hat{\mathcal{H}}(l, m, n)$ or just by $\hat{\mathcal{H}}$, if l, m, n are understood. (For further details about the construction of universal maps see [JS1] and for universal hypermaps see [CS].) It is not difficult to show that the hypermap $\mathcal{H}$ of type (l, m, n) can be identified with $\hat{\mathcal{H}}/\pi_1(\mathcal{H})$. This last statement is more correctly written as $\mathcal{H}$ is *isomorphic* to $\hat{\mathcal{H}}/\pi_1(\mathcal{H})$, where by an isomorphism between two hypermaps we just mean a homeomorphism between the underlying surfaces which maps brins to brins and preserves incidence. More generally we can define a morphism between hypermaps $\mathcal{H}_1$ and $\mathcal{H}_2$ to be a branched cover ϕ between the underlying surfaces, such that if b_2 is a brin of $\mathcal{H}_2$ then $\phi^{-1}(b_2)$ is a brin of $\mathcal{H}_1$, with a similar statement about the hypervertices, hyperedges and hyperfaces of $\mathcal{H}_2$. It may happen that a hypervertex, say, of valency n in $\mathcal{H}_1$ projects to a hypervertex of valency dividing n, if there is a branch point at the hypervertex (or center of the hypervertex, depending on which model we are using.) The same remark applies to hyperedges and hyperfaces. If there is a morphism between $\mathcal{H}_1$ and $\mathcal{H}_2$ then $\pi_1(\mathcal{H}_1)$ is conjugate to a subgroup of $\pi_1(\mathcal{H}_2)$, which is one reason for the terminology, 'fundamental group'. Another is that a hypermap has a natural orbifold structure and the fundamental group of the hypermap is the fundamental group of the orbifold. (See [Sc]).

5.7 Regular maps and hypermaps

If $\mathcal{H}$ is a hypermap with fundamental group $\pi_1(\mathcal{H}) < \Delta(l, m, n)$ then we say that $\mathcal{H}$ is regular if $\pi_1(\mathcal{H}) \triangleleft \Delta(l, m, n)$. This implies that the permutations g_0, g_1, g_∞ that generate the monodromy group in (5.1) are regular permutations, that is they are products of cycles of the same length. We take for these lengths the valencies of the hypervertices, hyperedges and hyperfaces. If, as is usual, we choose l, m, n as these lengths then $\pi_1(\mathcal{H})$ is torsion-free. A necessary and sufficient condition for a hypermap to be regular is that its automorphism group, (that is the group of sense-preserving homeomorphisms of the surface that keeps the hypermap invariant) is transitive on the brins of the hypermap. The most discussed case is that of regular maps which have been examined by many authors this century, notably Coxeter. For classical references on regular maps see [CM]. For regular hypermaps, see [CS, BJ]. For an example of a regular hypermap, see Fig.1(a). This is a hypermap with 9 brins and corresponds to a genus 1 regular hypermap with fundamental group a normal subgroup of $\Delta(3, 3, 3)$ isomorphic to $\mathbf{Z} \oplus \mathbf{Z}$.

5.8 Belyi functions and hypermaps

We saw above that that there is a correspondence between isomorphism classes of hypermaps of type (l, m, n) and conjugacy classes of subgroups of $\Delta(l, m, n)$, (namely the fundamental groups of the hypermaps). The trivial subgroup corresponds to the universal hypermap of type (l, m, n). The whole group $\Delta(l, m, n)$ corresponds to the trivial hypermap with just one brin (which may be regarded as the trivial map with one dart if $m = 2$.) We draw these trivial hypermaps in the Cori (fig. 2(a)) and James (fig. 2(b)) representation, and also the trivial map in Fig. 3. We see that the trivial hypermap has one hypervertex V_0, one hyperface F_0 and one hyperedge E_0. It also has one brin b_0. A similar remark applies to maps, except that now we have one (free) dart.

(a)

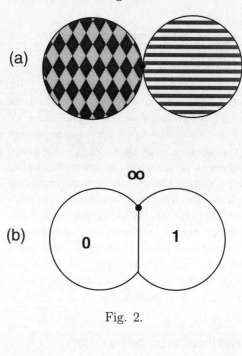

(b)

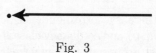

Fig. 2.

Fig. 3

Now let $\beta : \mathcal{R} \to \Sigma$ be a Belyi function. Then we can build a hypermap on $\mathcal{R}$ with hypervertices, hyperfaces and hyperedges $\beta^{-1}(V_0)$, $\beta^{-1}(F_0)$, $\beta^{-1}(E_0)$ respectively, and brins $\beta^{-1}(b_0)$. By Theorem 4.1, $\mathcal{R} \cong \mathcal{U}/\Gamma$ where Γ is a subgroup of a cocompact triangle group $\Delta(l, m, n)$. The hypermap then has type (l, m, n) and its fundamental group is Γ. Conversely, if we have a hypermap $\mathcal{H}$ of type (l, m, n) with fundamental group $\Gamma < \Delta(l, m, n)$ then it lies on the Riemann surface $\mathcal{R} = \mathcal{U}/\Gamma$ and there is a Belyi function $\mathcal{U}/\Gamma \longrightarrow \mathcal{U}/\Delta$. By Belyi's Theorm, $\mathcal{U}/\Gamma$ is the Riemann surface of an algebraic curve defined over $\overline{\mathbf{Q}}$. Only in a few cases can we easily identify this curve. One of these comes from Fig. 1. Here we obtain a torus corresponding to a normal subgroup of $\Delta(3, 3, 3)$. (This subgroup is the fundamental group of the hypermap.) We can show that this gives the Fermat cubic, which is $n = 3$ of example 2.3. The Belyi function is $\beta(x, y) = x^3$.

6   Smooth Belyi surfaces and Platonic surfaces

From the viewpoint of Riemann surfaces a major problem is the investigation of Belyi surfaces. These have the form $\mathcal{U}/\Gamma$ where Γ is a subgroup of finite index in a triangle group. The surfaces where we can choose Γ to be torsion-free (i.e. a surface group) are the most straightforward to study. Of these, the cases where additionally Γ is normal in the triangle group are of particular interest.

Definitions.

- 1. A Riemann surface is called a *smooth Belyi surface* if it is conformally equivalent to a surface of the form $\mathcal{U}/\Gamma$ where Γ is a torsion-free subgroup of finite index in a triangle group acting as a group of conformal automorphisms of $\mathcal{U}$, where $\mathcal{U}$ is either **H**, **C** or Σ.

- 2. A Riemann surface is called *platonic* if it is conformally equivalent to a surface of the form $\mathcal{U}/\Gamma$ where Γ is a torsion-free normal subgroup of finite index in a cocompact triangle group containing a period equal to 2. Such a surface carries a regular map.

- 3. A Riemann surface is called *quasiplatonic* if it is conformally equivalent to a surface of the form $\mathcal{U}/\Gamma$ where Γ is a torsion-free normal subgroup of finite index in a cocompact triangle group. Such a surface carries a regular hypermap. By our definition, $\mathcal{R}$ platonic implies that $\mathcal{R}$ is quasiplatonic which implies that $\mathcal{R}$ is a smooth Belyi surface.

The distinction between platonic and quasiplatonic is largely for historical reasons. In the literature on automorphism groups of Riemann surfaces and on maps or hypermaps, the platonic surfaces play an important rôle. In the work of some mathematicians (see for example [Br]) the word platonic is reserved for the case where the triangle group in 2 above has periods equal to 2 and to 3, (i.e. is an image of the modular group) but we prefer the slightly wider definition.

Example 6.1 *Genus zero. The Riemann sphere is platonic, as it contains the platonic solids! More precisely, the finite groups of rotations of the sphere are all finite triangle groups. We may regard the cyclic groups C_n as $\Delta(1, n, n)$, the dihedral groups D_n as $\Delta(2, n, n)$, and rotation groups of the tetrahedron, cube and dodecahedron as $\Delta(3, 2, 3)$, $\Delta(3, 2, 4)$ and $\Delta(3, 2, 5)$ respectively. The fundamental group of the corresponding maps is the identity subgroup which is normal.*

6.1   Smooth Belyi surfaces of genus one

The smooth Belyi surfaces must be of the form $\mathbf{C}/\Lambda$ where Λ is a lattice inside a Euclidean triangle group. There are only three Euclidean triangle groups up to conjugacy, these being $\Delta(2,4,4)$, $\Delta(2,3,6)$ and $\Delta(3,3,3)$. As $\Delta(3,3,3) < \Delta(2,3,6)$, we need not consider $\Delta(3,3,3)$. Now all lattices are similar to a lattice of the form $\Lambda_\tau = \mathbf{Z}(1,\tau)$, where τ has positive imaginary part.

The following is proved in [SS].

Theorem 6.2 $\Delta(2,4,4)$ *contains a lattice similar to* Λ_τ *if and only if* $\tau \in \mathbf{Q}(i)$. $\Delta(2,3,6)$ *contains a lattice similar to* Λ_τ *if and only if* $\tau \in \mathbf{Q}(\rho)$, *where* $\rho = \frac{-1+i\sqrt{3}}{2}$.

Now the set of τ in $\mathbf{H}$ form the Teichmüller space of the torus and Theorem 6.1 shows that the smooth Belyi surfaces are dense in this space. It is perhaps more important to consider the Modulus space in this case. As is well-known, (see chapter 6 of [JS2]) τ_1 and τ_2 in $\mathbf{H}$ represent the same conformal torus if and only if $\tau_2 = T(\tau_1)$, where $T \in G = PSL(2, \mathbf{Z})$. The modulus space of the torus is then obtained by identifying the sides of the standard fundamental region for G, namely $\{z = x + iy| -\frac{1}{2} \le x \le \frac{1}{2}, |z| \ge 1\}$, that is we identify the two vertical sides that are paired by $z \mapsto z + 1$ and the two halves of the segments of the the upper-half of the unit circle on either side of i, which are paired by $z \mapsto \frac{-1}{z}$. Thus the set of smooth Belyi surfaces of genus 1 form a dense subset of the genus 1 moduli space. Every conformal torus corresponds to an elliptic curve and by Belyi's Theorem, the genus 1 Belyi surfaces correspond to elliptic curves that may be defined over $\overline{\mathbf{Q}}$. In [SS], elliptic curves defined over $\mathbf{Q}$ or quadratic and cubic extensions of $\mathbf{Q}$ were found. For example, it is shown there that there are only 5 genus 1 Belyi surfaces that are defined over the rationals $\mathbf{Q}$.

6.2 Platonic surfaces of genus 1

It follows from §7 of [JS1] that there are only two platonic surfaces of genus 1, namely those that correspond to the points i and those that correspond to the point ρ of the fundamental region for G. For example, on a torus corresponding to i, there is the regular map $\hat{\mathcal{M}}(4,2,4)/\Lambda_i$, the quotient of the universal map of type $(4,2,4)$ by Λ_i. This is just the map consisting of one square face, one vertex and two edges. As $\Lambda_i \cong \mathbf{Z} \oplus \mathbf{Z}$, it contains infinitely many subgroups which are normal in $\Delta(2,4,4)$, and hence this torus carries infinitely many regular maps. These are called maps of type $\{4,4\}_{r,s}$ in [CM] and are also discussed in §7 of [JS1]. Similar remarks correspond to Λ_ρ. As every lattice that is normal in $\Delta(3,3,3)$ is also normal in $\Delta(2,3,6)$ it follows that every quasiplatonic surface of genus 1 is platonic.

6.3 Smooth Belyi surfaces of genus 2 and higher genus

A smooth Belyi surface of genus $g > 1$ corresponds to a torsion-free Fuchsian group of genus g inside a cocompact triangle group. The index of the inclusion is finite, and so we find only finitely many such triangle groups are possible, and each triangle group can contain only finitely many genus g torsion-free subgroups. Thus there are only finitely many smooth Belyi surfaces of genus g. In some cases, two different hypermaps will give rise to the same Riemann surface. For example, a map and its dual give the same Riemann surface. However, as we shall discuss, there are more subtle ways where two distinct hypermaps can give the same Riemann surface. (From [Sy], table 9 it follows that there are 581 uniform hypermaps of genus 2, where we don't regard a hypermap obtained just by permuting hypervertices, hyperedges and hyperfaces as being different. Thus there are at most 581 smooth Belyi surfaces of genus 2.)

6.4 Platonic surfaces of genus 2 and higher genus

A platonic surface of genus 2 comes from a normal torsion-free genus 2 subgroup of a triangle group $\Delta(m, 2, n)$ and so corresponds to a regular map of type (m, n) on a surface of genus 2. (This means that every vertex has valency m and every face has valency n.) Up to duality there are 6 regular maps on a surface of genus 2, of types $(8, 8)$, $(5, 10)$, $(6, 6)$, $(4, 8)$, $(4, 6)$, $(3, 8)$, ([CM], table 9). However, we have inclusion relationships between triangle groups $\Delta(8, 2, 8) < \Delta(4, 2, 8) < \Delta(3, 2, 8)$, $\Delta(6, 2, 6) < \Delta(4, 2, 6)$. Now $\Delta(8, 2, 8)$ contains a normal torsion-free genus 2 subgroup of index 8 and we can show that this subgroup is also normal in $\Delta(3, 2, 8)$ and $\Delta(4, 2, 8)$. Thus we have the same platonic surface with 3 distinct regular maps. A similar remark applies to the inclusion $\Delta(6, 2, 6) < \Delta(4, 2, 6)$, so that there are only 3 platonic surfaces of genus 2. These have automorphism groups of orders 10, 24, and 48. In a similar way, we find that there are 7 platonic surfaces of genus 3, 7 platonic surfaces of genus 4 and 9 platonic surfaces of genus 5, see [SW].

6.5 Quasiplatonic surfaces of genus 2 and higher genus

It may be shown that all genus 2 quasiplatonic surfaces are platonic. This follows by the detailed analysis of genus 2 regular hypermaps in [CS, BJ]. However in genus 3, we can find a quasiplatonic, but not platonic surface as follows. There is a homomorphism $\theta : \Delta(3, 9, 9) \to C_9$, whose kernel is a torsion-free genus 3 group. There is an index two inclusion $\Delta(3, 9, 9) < \Delta(2, 6, 9)$. It can be shown that the homomorphism $\theta : \Delta(3, 9, 9) \to C_9$ does not extend to a homomorphism θ from $\Delta(2, 6, 9)$ to a group of order 18. (In fact according to Broughton's list ([Bro], p255) there is no action of a group of order 18 on a surface of genus 3.)

7   Hypermaps versus Riemann surfaces

We have seen that the existence of a hypermap $\mathcal{H}$ on a surface X implies that X can be made into a Riemann surface. Precisely, a hypermap corresponds to a subgroup Γ of some triangle group Δ and the Riemann surface is then $\mathcal{U}/\Gamma$. This gives a function $F : \mathbf{Hyp}(g) \mapsto \mathbf{M}(g)$, where $\mathbf{Hyp}(g)$ is the set of hypermaps of genus g and $\mathbf{M}(g)$ is the moduli space of Riemann surfaces of genus g. This function is certainly not surjective if $g > 0$ as the image, by definition, consists of Belyi surfaces and these correspond to algebraic curves defined over $\overline{\mathbf{Q}}$. Also it is not injective either, (as we shall see) so that a compact Belyi surface will usually correspond to more than one hypermap. We now restrict F to the set $\mathbf{RM}$ of regular maps of genus g (where we identify a map with its dual) and consider the injectivity of F on $\mathbf{RM}$. The image of $\mathbf{RM}$ consists of the set of Platonic surfaces. We have already seen examples where this function is not injective. For genus 0 there are infinitely many regular maps on the sphere (if one allows maps of type $(2, n)$) so for $g = 0$ F is infinitely many to one. Also for $g = 1$, we saw that the platonic surfaces correspond to the points i and ρ, which carry regular maps of types $(4, 4)$ and $(3, 6)$ respectively. We saw above that each such surface carries infinitely many regular maps, as a lattice will contain infinitely many other lattices normal in the triangle group. We will see a more geometric reason shortly. The typical situation occurs when $g = 2$. We saw above that we obtain distinct regular maps on the same Riemann surface in the case when we have inclusion relationships between triangle groups containing one period equal to 2. We can list all such inclusion relationships from the lists given in [S1]. They are as follows.

- (i) $\Delta(n, 2, n) < \Delta(4, 2, n)$ with index 2.

- (ii) $\Delta(2n, 2, n) < \Delta(3, 2, 2n)$ with index 3.

- (iii) $\Delta(8, 2, 8) < \Delta(3, 2, 8)$ with index 6.

- (iv) $\Delta(7, 2, 7) < \Delta(3, 2, 7)$ with index 9.

The first two inclusion relationships hold for all integers $n \geq 3$, and we will describe them geometrically. The final two inclusion relationships are more special but they do occur in some important examples in low genus as we shall point out.

The inclusion $\Delta(n, 2, n) < \Delta(4, 2, n)$. This gives the *medial map*. If we have a regular map $\mathcal{M}$ of type (n, n) then we build a new map med$(\mathcal{M})$ whose vertices lie at the midpoints of the edges of $\mathcal{M}$, and whose edges are the *medials* of $\mathcal{M}$, that is edges that join the midpoints of adjacent edges of $\mathcal{M}$ that are consecutive around a face. The map $\mathcal{M}$ has two kinds of faces. There are faces of size n whose face centres are at the face centres of $\mathcal{M}$ and faces of size n whose face centres are at the vertices of $\mathcal{M}$. If we tried this construction with a map of type (m, n) then the medial map would have faces of sizes m and n and so would not be regular if $m \neq n$. The medial map is not necessarily regular if $m = n$, but there are many examples when it is. An example of medial maps is obtained by considering the tetrahedral map of type $(3,3)$ on the sphere when the medial map is the octahedron. When $n = 4$ and $g = 1$ we get a map of type $(4, 4)$ on the torus and the medial map is also of type $(4, 4)$. (In this way we get infinitely many regular maps of type $(4, 4)$ on the platonic torus correponding to i.)

The inclusion relationship $\Delta(2n, 2, n) < \Delta(3, 2, 2n)$. This gives the *truncation* of $\mathcal{M}$, trunc($\mathcal{M}$). Following [Sy], we construct the truncation by placing two vertices of trunc($\mathcal{M}$) on each edge of $\mathcal{M}$. We form edges of trunc($\mathcal{M}$) as follows. One edge joins the two new vertices of trunc($\mathcal{M}$) on the same edge of $\mathcal{M}$ by following along this edge. Consider the two vertices of trunc($\mathcal{M}$) on the same edge of $\mathcal{M}$. If we call these v_1 and v_2 then the edge of $\mathcal{M}$ from v_2 to v_1 will meet a vertex V of $\mathcal{M}$ (and we say that v_1 is *closer* to V than v_2). If we rotate this edge of $\mathcal{M}$ clockwise and anticlockwise about V we arrive at two other edges of $\mathcal{M}$. Each of these edges contains a vertex of trunc($\mathcal{M}$) closer to V in the above sense. We join v_1 to each of these vertices. As V has valency $2n$ by repeating this process we obtain a $2n$-gon in trunc($\mathcal{M}$) with centre V. We now see that the vertices of trunc($\mathcal{M}$) have valency 3 while the faces are all $2n$-gons. For example if $n = 3$ we find a truncation of a regular map of type $(6,3)$ on a torus is a another regular map of type $(3,6)$ on the same torus and in this way we find that the platonic surface corresponding to ρ contains infinitely many regular maps.

In both the cases we have discussed above the regular map $\mathcal{M}$ of type (n, n), (resp. $(2n, n)$) is defined by its fundamental group $\pi_1(\mathcal{M})$ inside the triangle group $\Delta(n, 2, n)$ (resp. $\Delta(2n, 2, n)$). It can readily be shown using fundamental regions that the medial map (resp. truncation) is defined by their fundamental groups in $\Delta(4, 2, n)$ (resp. $\Delta(3, 2, 2n)$). (The details are given in [Sy]). This means that in both cases the medial map and the truncation both lie on the same Riemann surface as the original map, (of type (n, n) in the first case, or of type $(2n, n)$ in the second case). In terms of regular maps the inclusion relationships (iii) and (iv) are less geometric and probably arise much less often. However there are two important cases that deserve mention. In genus 2, the maximum order of a group of automorphisms of a Riemann surface is 48 and there is a unique Riemann surface X_{48} with this property. This correponds to the group $GL(2, 3)$ as an image of $\Delta(3, 2, 8)$. The inclusion relationships $(2; -) < \Delta(8, 2, 8) < \Delta(4, 2, 8) < \Delta(3, 2, 8)$ show that in this case the relationship between the maps of type $(8,8)$ and $(3,8)$ is just take the

medial map then consider the dual and then consider the truncation. This
shows that X_{48} carries 3 regular maps.

The inclusion relationship $\Delta(7, 2, 7) < \Delta(3, 2, 7)$ is important in genus 7. Images of $\Delta(3, 2, 7)$ are important as they correspond to Riemann surfaces admitting $84(g-1)$ automorphisms, the greatest number by Hurwitz's Theorem.
For $g = 7$ the corresponding Hurwitz group is $PSL(2, 8)$ of order 504. This
contains an affine subgroup of order 56, corresponding to the triangle group
$\Delta(7, 2, 7)$, (see [M] for a discussion from the point of view of algebraic curves.)
The associated regular map of type $(7,7)$ is obtained from an embedding of
the complete graph K_8 in a surface of genus 7. This gives the first example of
a chiral map of genus $g > 1$, due to Edmonds. (See [Cox] and [S2]).

Given a map $\mathcal{M}$, then its dual map is obtained by interchanging face centres
by vertices. More precisely, we place a vertex at the centre of each face and
join two such vertices by an edge if the corresonding faces are adjacent. If a
map has type (m, n), then its dual has type (n, m). Furthermore, the dual of a
regular map is regular. Algebraically, we replace the triangle group $\Delta(2, m, n)$
by the isomorphic triangle group $\Delta(2, n, m)$.

Theorem 7.1 *Let $\mathcal{M}_1$ and $\mathcal{M}_2$ be two regular maps of genus $g > 1$ and
suppose that $F(\mathcal{M}_1) = F(\mathcal{M}_2)$, where $F : \mathbf{RM} \mapsto \mathbf{M}(g)$ is the function from
the set of regular maps to the moduli space of genus g described above. We
also suppose that $E(\mathcal{M}_1) \le E(\mathcal{M}_2)$ where $E(\mathcal{M})$ denotes the number of edges
of $\mathcal{M}$. Then either*

*(a) $\mathcal{M}_1$ has type (n, n) and $\mathcal{M}_2$ is the medial map or the dual of the medial
map of $\mathcal{M}_1$,*

*(b) $\mathcal{M}_1$ has type $(n, 2n)$ and $\mathcal{M}_2$ is the truncation or the dual of the trucation
of $\mathcal{M}_1$,*

(c) $\mathcal{M}_1$ has type $(8,8)$ and $\mathcal{M}_2$ has type $(3,8)$ or $(8,3)$,

(d) $\mathcal{M}_1$ has type $(7,7)$ and $\mathcal{M}_2$ has type $(3,7)$ or $(7,3)$,

(e) $\mathcal{M}_1 \cong \mathcal{M}_2$ up to duality.

Corollary 7.2 *The function $F : \mathbf{RM} \mapsto \mathbf{M}(g)$ is injective on the set of
regular maps of type (m, n) if $m \notin \{n, 2n, n/2\}$.*

Proof of Theorem. Suppose that $F(\mathcal{M}_1) = F(\mathcal{M}_2)$. Then $\mathbf{H}/M_1 \cong \mathbf{H}/M_2$ and because M_1 and M_2 are surface groups it follows that they are conjugate in $PSL(2, \mathbf{R})$, and thus $M_2 = g M_1 g^{-1}$, $(g \in PSL(2, \mathbf{R}))$. As the maps $\mathcal{M}_1$ and $\mathcal{M}_2$ are regular $M_1 \triangleleft \Delta(m_1, 2, n_1)$ and $M_2 \triangleleft \Delta(m_2, 2, n_2)$. Also, $M_1 \triangleleft g^{-1}\Delta(m_2, 2, n_2)g$. There are two cases:

- (1) $g^{-1}\Delta(m_2, 2, n_2)g \neq \Delta_1(m_1, 2, n_1)$, and

- (2) $g^{-1}\Delta(m_2, 2, n_2)g = \Delta_1(m_1, 2, n_1)$.

In (1) the normalizer of M_1 in $PSL(2, \mathbf{R})$ properly contains the triangle group $\Delta(m_1, 2, n_1)$. As the normalizer of a non-cyclic Fuchsian group is Fuchsian and as the only Fuchsian group that properly contains a triangle group is a triangle group we are in one of the cases (i), (ii), (iii), (iv) mentioned above. As we have seen, these correspond to cases (a), (b), (c), (d) of the Theorem. If (2) holds, then $m_1 = m_2$, $n_1 = n_2$ (or the dual case $m_1 = n_2$, $m_2 = n_1$) and so $E(\mathcal{M}_1) = E(\mathcal{M}_2)$. As the structure of the map just depends on the permutation representation of the cosets of $\pi_1(\mathcal{M})$ in the triangle group, it follows that $\mathcal{M}_1 \cong \mathcal{M}_2$ or its dual.

8 Conclusion

The theory of maps and hypermaps on surfaces, also known as dessins d'enfants has been developed largely with applications to Galois Theory in mind. For a survey of such material see [Jo]. For recent papers on the subject the three books of lectures published in the London Mathematical Society Lecture Note series in volumes 200, 242 and 243 edited by Schneps or Schneps and Lochak, (see [SL]) are a good source. In particular, volume 242 contains Grothendieck's *Esquisse d'un programme* both in the original and with an English translation. However, our main interest in this article, namely the connections between maps, and especially regular maps with Riemann surfaces goes back to Felix Klein (see [L], which contains Klein's original paper on his Riemann surface of genus 3, together with commentaries and other papers related to this surface.) For more recent work the reader is referred to [Br], [JS1], [JS3], [KW], [MS], [S2], [S3], [S4], [SSy], [SW], [Wo]. Connections with symmetric Riemann surfaces and real curves are described in [S1] and [MS] with hyperelliptic surfaces in [S4], and with elliptic curves in [SSy]. For relations with the geometric structure and in particular the differential structure of the surface see [Br] and [KW]. In [SW] we study Weierstrass points on Platonic surfaces in relation to the structure of the underlying regular map. On a Riemann surface of genus $g \geq 2$ there is a collection of W Weierstrass points where $2g+2 \leq W \leq g^3 - g$. The Weierstrass points are left invariant by the automorphism group of the

Riemann surface, and hence by the automorphism group of the underlying regular map. A natural question is then to ask whether the Weierstrass points are vertices, face-centres or edge-centres of the map, and the answer is affirmative in low genus, but not always for higher genus. A well-known example is the Klein map. This has 24 Weierstrass points and they are all at the centres of the 24 heptagons on the surface.

I would like to thank Gareth Jones for his discussions over the years on this topic, and also my research students Paul Watson, Robert Syddall, Adnan Melekoğlu and Ioannis Ivrissimtzis for working with me on dessins and helping me understand many of the ideas presented here. Thanks also to Pablo Martín and Gareth Jones for reading a preliminary draft of this article and suggesting some improvements.

References

[B] G. V. Belyi, 'On Galois extensions of a maximal cyclotomic field', *Math. USSR-Izv. 14* (1980) 247-256 (English translation).

[BJ] A. J. Breda d'Azevedo and G. A. Jones, 'Rotary hypermaps of genus 2,' *Contributions to Algebra an Geometry* (to appear.)

[Br] R. Brooks, 'Platonic surfaces', *Comment. Math.Helv. 74* (1999), 156-170.

[Bro] S. A. Broughton,'Classifying finite group actions on surfaces of low genus' *Journal of pure and applied algebra 69)* (1990),233-70.

[CIW] P. B. Cohen, C. Itzykson and J. Wolfart, 'Fuchsian triangle groups and Grothendieck dessins. Variations on a theme of Belyi' *Comm. Math. Physics. 163* (1994),605-627.

[C] R. Cori, 'Un Code pour les graphes planaires et ses applications', *Astérisque 27* (1975)

[CS] D.Corn and D.Singerman,'Regular hypermaps',*European J. Combinatorics, 9* (1988) 337-351.

[Cox] H. S. M. Coxeter, *'Introduction to Geometry'*, John Wiley and Sons, 1962.

[CoxM] H. S. M. Coxeter and W. O. J. Moser, *'Generators and relations for discrete groups'* (Springer,Berlin 1965).

[FK] H. M. Farkas and I. Kra, *'Riemann surfaces'* Graduate Texts in Mathematics, No. 81, Springer-Verlag, (1981).

[G] P. A. Griffiths, *'Introduction to algebraic curves'*, Translation of Mathematical Monographs, vol 76, American Mathematical Society (1989).

[J] L. D. James, 'Operations on hypermaps and outer automorphisms', *European J. Combinatorics, 9* (1988) 551-560.

[Jo] G. A. Jones, 'Maps on surfaces and Galois groups',*Math. Slovaca, 47* (1997) 1-33.

[JS1] G. A. Jones and D. Singerman, 'Theory of maps on orientable surfaces', *Proc. London Math. Soc.(3) 37* 273-307.

[JS2] G. A. Jones and D. Singerman,*'Complex functions, an algebraic and geometric viewpoint'* (Cambridge University Press, 1987)

[JS3] G. A. Jones and D. Singerman, 'Belyi functions, hypermaps and Galois groups', Bull. London Math. Soc 28 (1996) 561-590

[KW] H. Karcher and M.Weber, 'The geometry of Klein's Riemann surface,' in Silvio Levy, (ed) *The Eightfold Way* MSRI research publications, Cambridge University Press, (1999).

[L] S. Levy (ed) *The Eightfold Way*, MSRI research publications, Cambridge University Press (1999)

[M] A. M. Macbeath, 'On a curve of genus 7', *Proc. London Math. Soc 15.* (1965) 527-542.

[Mag] W. Magnus, *'Non-Euclidean tesselations and their groups'.* Academic press, 1974.

[MS] A. Melekoğlu and D. Singerman, 'Reflections of regular maps and Riemann surfaces,' preprint, University of Southampton.

[Sc] P. Scott, 'The Geometries of 3-manifolds', *Bull. London Math.Soc. 15*(1983) 401-447.

[SL] L. Schneps and P. Lochak (eds) *Geometric Galois actions 1* London Mathematical Society Lecture Note Series, 242 Cambridge University Press, 1997.

[SZ] G, Shabat and A. Zvonkin, 'Plane trees and algebraic numbers', *Contemporary Mathematics, 178,* (1994) 233-275.

[S1] D. Singerman, 'Finitely Maximal Fuchsian groups', *J. London Math.Soc. (2) 6* (1972) 29-38.

[S2] D. Singerman, 'Symmetries of Riemann surfaces with large automorphism group', *Math. Annalen 210* 17-32 (1974)

[S3] D. Singerman, 'Klein's Riemann surface of genus 3 and regular imbeddings of finite projective planes', *Bull. London Math. Soc. 18* (1986), 364-370.

[S4] D. Singerman, 'Hyperelliptic maps and surfaces', *Math. Slovaca 47*(1997) 93-97.

[SSy] D. Singerman and R. I. Syddall, 'Belyi uniformization of elliptic curves', *Bull. London Math. Soc. 29* (1997) 443-451.

[SW] D. Singerman and P. Watson, 'Weierstrass points on regular maps', *Geom. Dedicata 66,* (1997), 69-88.

[Sy] R. I. Syddall, *'Uniform dessins of low genus'*, University of Southampton Ph.D.thesis, (1997).

[W] A. T. White, *Graphs, groups and surfaces* Revised edition, North-Holland, (1984)

[Wo] J. Wolfart, 'The 'obvious' part of Belyi's theorem and Riemann surfaces with many automorphisms.' in L. Schneps and P. Lochak (eds) *Geometric Galois actions 1* London Mahematical Society Lecture Note Series, 242 Cambridge University Press, (1997.) 97-112.

Faculty of Mathematical Studies,

University of Southampton,

Southampton SO17 1BJ.

COMPACT RIEMANN SURFACES

AND ALGEBRAIC FUNCTION FIELDS

Peter Turbek

This paper is divided into two parts. In the first part we prove that the category of compact Riemann surfaces is functorially equivalent to the category of algebraic function fields of one variable over $\mathbb{C}$. In the second part we consider a specific Riemann surface defined by an algebraic equation. We illustrate computational techniques which are employed to determine the genus, holomorphic differentials, and full automorphism group of the surface.

The material contained in the first part is standard. The reader will recognize the influence of Farkas-Kra [4], Fulton [5], Springer [7], and especially Chevalley [3]. Although the techniques demonstrated in the second part of the paper are known, I know of few examples in the literature where such a detailed analysis of a particular Riemann surface is given.

1 Meromorphic Function Fields and Algebraic Function Fields

Throughout the paper all Riemann surfaces considered are compact and all fields considered are extensions of the complex numbers $\mathbb{C}$. We denote the Riemann sphere by $\hat{\mathbb{C}}$. An algebraic function field of one variable over $\mathbb{C}$ is a field extension K of transcendence degree one over $\mathbb{C}$. This means that there is an element $x \in K$ such that x is transcendental over $\mathbb{C}$ and K is a finite field extension of $\mathbb{C}(x)$. Let X be a compact Riemann surface. It is well known that X admits nonconstant meromorphic functions and it is easy to see that the set of meromorphic functions of X forms a field $\mathcal{M}(X)$. We will prove that $\mathcal{M}(X)$ is an algebraic function field of one variable. Conversely, we will prove that to each algebraic function field of one variable K, we can associate a compact Riemann surface $\mathcal{S}(K)$. This correspondence is accomplished by noting that each point of X yields an algebraic structure, namely, a discrete valuation ring contained in $\mathcal{M}(X)$. We will show that each discrete valuation ring in $\mathcal{M}(X)$ arises from a point of X, and distinct points of X yield distinct valuation rings. Thus the points of X are in one to one correspondence with the set of discrete valuation rings in $\mathcal{M}(X)$. If K is an algebraic function field, we exploit this idea to define a topology on the discrete valuation rings contained in K so that the resulting object is a Riemann surface.

1.1   Valuations

Let X be a compact Riemann surface. Since X is a complex manifold, at each point $p \in X$ we can choose a neighborhood U of p and a function $z = f(q)$ such that the map $f : U \to \mathbb{C}$ maps U homeomorphically onto a neighborhood of $z = 0$ in the complex plane. Locally, every meromorphic function of X can be expressed in terms of z and has a Laurent series at p. If $g \in \mathcal{M}(X)$ and locally g can be expressed as $g(z) = z^k h(z)$, where $h(z)$ is holomorphic at $z = 0$ and does not vanish there, we define $ord_p(g) = k$. It is easy to see that ord_p does not depend on the choice of local coordinates. Then ord_p is a valuation on $\mathcal{M}(X)$, and associated to it are several algebraic structures.

Definition 1 *Let K be a field defined over $\mathbb{C}$, and let K^* denote the nonzero elements of K. Let $v : K^* \to Z$ be a surjective map which, for all f and g in K^* satisfies:*

1. $v(fg) = v(f) + f(g)$

2. $v(f + g) \geq min\{v(f), v(g)\}$.

Then we say v is a discrete valuation on K. We may extend v to all of K by defining $v(0) = \infty$. With this definition, given the usual conventions of adding integers and ∞, i) and ii) still hold.

Proposition 1 *Let K be a field and let v be a discrete valuation on K. Define $R := \{f \in K \mid v(f) \geq 0\}$ and let $I := \{f \in K \mid v(f) > 0\}$. Then R is an integral domain and I is the unique maximal ideal of R. In addition, I is principally generated.*

Proof: Since K is a field, R inherits properties such as the distributive laws from K. Note that $0 \in R$ by definition. From property 1), we have $v(1) = v((1)(1)) = v(1) + v(1)$. This immediately yields that $v(1) = 0$, thus $1 \in R$. A similar argument shows $v(-1) = 0$. Properties 1) and 2) show respectively, that R is closed under multiplication and addition, thus R is a ring. Property 2) also yields that I is closed under addition. If $f \in R$ and $g \in I$, then $v(f) \geq 0$ and $v(g) > 0$, thus Property 2) yields that $v(fg) > 0$, thus $fg \in I$, showing that I is an ideal. Since K is a field, R is an integral domain. Note that if $g \in R$, then $v(1/g) = -v(g)$, since $0 = v(1) = v(g/g) = v(g) + v(1/g)$. We now show that I is principally generated. Let t be any element with $v(t) = 1$. Let $h \in I$. Then $v(h) > 0$, say $v(h) = k$. Then $v(h/t^k) = v(h) - v(t^k) = 0$. Thus $r := h/t^k \in R$. But this yields that $rt^k = h$, thus $h \in (t)$, the ideal generated by t. Thus $(t) = I$. We now show that I is maximal. If $g \in R \setminus I$, then $v(g) = 0$ and $v(1/g) = -v(g) = 0$. Therefore $1/g \in R$, thus R is a unit. Therefore every element not in I is a unit. Thus I is maximal.

To each valuation v we have associated an integral domain R with a unique maximal ideal which is principally generated. A ring which arises in this manner is called a discrete valuation ring. Conversely, to each Noetherian integral domain R with unique maximal ideal I which is principally generated, a valuation can be associated to its field of fractions K. The valuation ord_I can be easily described. If x generates I and if $r \in R$, with $r = x^k y$ where $y \in R \setminus I$, then we define $ord_I(r) = k$. This clearly extends to K by defining $ord_I(r/s) = ord_I(r) - ord_I(s)$ for all r and s in K. We shall state, without proof, an equivalent characterization of discrete valuation rings found in [3].

Proposition 2 *Let K be an algebraic function field of one variable over $\mathbb{C}$. Let R be a proper subring of K which contains $\mathbb{C}$. Assume that for all $x \in K$, if $x \notin R$ then $1/x \in R$. Then R is a discrete valuation ring.*

Although we will not prove Proposition 2, we now verify that all discrete valuation rings contain $\mathbb{C}$.

Proposition 3 *Let K be an algebraic function field of one variable over $\mathbb{C}$ and let R be a discrete valuation ring of K. Then R contains $\mathbb{C}$.*

Proof: Let v be the discrete valuation associated to R. Let c be any complex number. We will show that $v(c) = 0$. If $v(c) = n \neq 0$, then choose $d \in \mathbb{C}$ such that $d^{n+1} = c$. Then $n = v(c) = v(d^{n+1}) = (n+1)v(d)$, a contradiction. Thus $v(c) = 0$. Thus $\mathbb{C}$ is contained in R.

For later reference we state three properties of valuations.

Proposition 4 *Let v be a valuation on a field K, and let f and g be elements of K. If $v(f) < v(g)$, then $v(f + g) = v(f)$.*

Proof: Let R be the ring $R = \{h \in K \mid v(h) \geq 0\}$, and let I be the ideal consisting of those elements $h \in R$ such that $v(h) > 0$. Assume $v(f+g) > v(f)$. Since $f \neq 0$, $v(f + g) = v(f(1 + g/f)) = v(f) + v(1 + g/f)$. This implies that $v(1 + g/f) > 0$, hence $1 + g/f \in I$, however, since $v(f) < v(g)$, $v(g/f) > 0$, hence $g/f \in I$ also. Thus we conclude that $1 \in I$, a contradiction.

Note that by induction, we may extend this result to the following. Let $v(f) < v(f_i)$ for $i = 1, 2, \ldots, n$. Then $v(f + f_1 + \ldots + f_n) = v(f)$.

Proposition 5 *Let v be a valuation on a field K. Suppose $t^n + a_{n-1}t^{n-1} + \ldots + a_1 t + a_0 = 0$, for elements $a_i, t \in K$. If $v(t) < 0$, then for some i, $v(a_i) < 0$.*

Proof: If $v(t) < 0$, then $v(t^n) < v(t^k)$ for $k = 1, \ldots, n-1$. If $v(a_i) \geq 0$ for each i, then the previous result implies that $v(t^n + a_{n-1}t^{n-1} + \ldots + a_1 t + a_0) = v(t^n)$, contradicting that $t^n + a_{n-1}t^{n-1} + \ldots + a_1 t + a_0 = 0$.

Proposition 6 *Let v be a valuation on K. Assume $z \in K$ and $v(z) = 1$. Suppose $z^k r(z)/s(z)$ is a nonzero rational expression in z where z does not divide $r(z)$ or $s(z)$. Then $v(z^k r(z)/s(z)) = k$.*

Proof: Clearly $v(z^k) = k$. Since $\mathbb{C}$ is algebraically closed, we may express $r(z)$ as $a_0(z - a_1) \ldots (z - a_n)$, and $s(z) = b_0(z - b_1) \ldots (z - b_m)$ where no a_i or b_i is zero. Then $v(r(z)/s(z)) = v(z-a_1) + \ldots + v(z-a_n) - v(z-b_1) - \ldots - v(z-b_m)$. From Proposition 4, $v(z - a_i) = 0 = v(z - b_j)$ for $1 \leq i \leq n$, and $1 \leq j \leq m$. Thus $v(z^k r(z)/s(z)) = k$.

Let X be a Riemann surface and let p be a point of X. We defined ord_p at the beginning of this section. It is easy to see that ord_p is a valuation. We define O_p to be the set of meromorphic functions of X which are defined at p. It is a discrete valuation ring with maximal ideal I_p, the set of functions which vanish at p. Any meromorphic function $f \in \mathcal{M}(X)$ with $ord_p(f) = 1$ generates I_p. Therefore each point p of X yields a valuation ord_p and an associated discrete valuation ring O_p.

1.2   Riemann Roch Theorem

We now recall the Riemann Roch theorem and some of its consequences. Let X be a compact Riemann surface. Recall that the group of divisors of X is the free abelian group generated by the points of X. Therefore, if D is a divisor, then D is a finite sum $D = \sum_{i=1}^{r} n_i P_i$ where each n_i is an integer and each P_i is a point of X. We define $deg(D) := n_1 + n_2 + \ldots + n_r$. We say $D \geq 0$ if each $n_i \geq 0$. To each $f \in \mathcal{M}(X)$, we associate a divisor, denoted by (f) as follows. Recall that since X is compact, f has a finite number of zeros, say $P_1, P_2, \ldots, P_r$ and a finite number of poles, say $Q_1, \ldots, Q_s$. We define $(f) = \sum_{i=1}^{r} n_i P_i + \sum_{j=1}^{s} n_j Q_j$ where $n_i = ord_{P_i}(f)$ and $n_j = ord_{Q_j}(f)$. We define the zero divisor of f, denoted by $(f)_0$, by $(f)_0 = \sum_{i=1}^{r} n_i P_i$ and we define the pole divisor of f by $(f)_\infty := -\sum_{j=1}^{s} n_j Q_j$. Note that $(f) = (f)_0 - (f)_\infty$ and that $(f)_0 \geq 0$ and $(f)_\infty \geq 0$. It is well known that $deg(f) = 0$ for each nonconstant $f \in \mathcal{M}(X)$. We can associate a divisor to each meromorphic differential ω on X. Locally at $p \in X$, ω can be expressed as $\omega = h(z)dz$, where $z \in \mathcal{M}(X)$ and $ord_p(z) = 1$. We define $ord_p(\omega) = ord_p(h)$ and obtain a divisor by associating $ord_p(\omega)$ to each point p of X. Two divisors D and D' are said to be equivalent if there exists $f \in \mathcal{M}(X)$ such that $D' = D + (f)$. A canonical divisor is any divisor equivalent to the divisor of a meromorphic differential. If D is a divisor, we define $L(D) := \{f \in \mathcal{M}(X) \mid (f) + D \geq 0\}$. It is easy to see that $L(D)$ is a finite dimensional vector space over $\mathbb{C}$. We define $l(D) := dim(L(D))$, the dimension of $L(D)$ as a vector space over $\mathbb{C}$. The Riemann Roch Theorem states the following:

Theorem 7 (Riemann Roch) *Let X be a compact Riemann surface of genus g, let D be any divisor, and let W be a canonical divisor. Then*

$$l(D) = deg(D) + 1 - g + l(W - D).$$

We have the following corollaries. Each corollary continues to use the notation of the Riemann Roch theorem.

Corollary 8 $l(W) = g$.

Proof: In the Riemann Roch theorem, let $D = 0$. We obtain, $l(0) = 0 + 1 - g + l(W)$. However if $g \in L(0)$, then g is defined everywhere on X, thus g is a constant function. Thus $l(0) = 1$. Thus $l(W) = g$.

Corollary 9 $deg(W) = 2g - 2$.

Proof: In the Riemann Roch theorem, let $D = W$. Then $l(W) = deg(W) + 1 - g + l(W - W)$. From above we know that $l(0) = 1$ and $l(W) = g$. Substituting these values in yields $deg(W) = 2g - 2$.

Corollary 10 *If $deg(D) > 2g - 2$, and if P is any point of X, then there exists an $f \in \mathcal{M}(X)$ such that $f \in L(D + P)$ and $f \notin L(D)$.*

Proof: Since $deg(D) > 2g - 2$, $deg(W - D) < 0$, thus a function in $L(W - D)$ would have to have more zeros than poles. Thus $l(W - D) = 0$, so $l(D) = deg(D) + 1 - g$, while $l(D + P) = deg(D) + 1 + 1 - g = 1 + l(D)$.

If P is distinct from the support of D, then f in Corollary 10 has a simple pole at P.

Corollary 11 *Let X be a compact Riemann surface, let $P_1, P_2, \ldots, P_n$ be distinct points of X, and let $b_1, \ldots, b_n$ be distinct complex numbers. Then there is an $f \in \mathcal{M}(X)$ such that $f(P_i) = b_i$.*

Proof: Pick $Q \notin \{P_1, P_2, \ldots, P_n\}$ and let D denote the divisor $(2g - 1)Q$. From the above, $l(D) = g$, and, from Corollary 10, for each i, there is a function f_i such that $f_i \in L(D + P_i)$ and $f_i \notin L(D)$. Each f_i necessarily has a pole of order one at P_i, and by adding a constant function to it, if necessary, we may assume it does not have a zero at any other P_i. Let $F = 1/(f_1 f_2 \ldots f_n)$, let $F_i = F f_i$, and let $a_i = F_i(P_i)$. Note that F has a simple zero at each P_i (as well as a zero at Q if the genus of X is not 0), that $F_i(P_j) = 0$ if $j \neq i$, and that $F_i(P_i) = a_i \neq 0$. Then the function

$$G := \sum_{i=1}^{n} \frac{b_i}{a_i} F_i$$

has the property that $G(P_i) = b_i$.

1.3 The field of meromorphic functions of a Riemann surface

Let X be a compact Riemann surface of genus g. We will show that $\mathcal{M}(X)$ is an algebraic function field of one variable over $\mathbb{C}$.

It is easy to see that each nonconstant function z is transcendental over $\mathbb{C}$. If not, then $z \in \mathcal{M}(X) \setminus \mathbb{C}$ satisfies a polynomial $G(T) := T^n + a_1 T^{n-1} + \ldots + a_{n-1}T + a_0 = 0$ with coefficients in $\mathbb{C}$. Since $\mathbb{C}$ is algebraically closed, $G(T)$ factors over $\mathbb{C}$, say $G(T) = (T - b_1)(T - b_2)\ldots(T - b_n) = 0$. Thus z is the constant function $z = b_i$, for some i.

Definition 2 *Let f be a holomorphic function in a neighborhood U of $z = 0$ of the complex plane. We say f is k to one on U if for each nonzero $z \in U$, there are exactly k distinct points of U, $z_1 = z, z_2, \ldots, z_k$, such that $f(z_i) = f(z)$.*

Proposition 12 *Let $f = z^k h(z)$ be a holomorphic function with $h(0) \neq 0$. Then there is a neighborhood U of $z = 0$ on which f is k to one.*

Proof: Since $h(0) \neq 0$, U contains a small disc, centered at 0, in which $h(z) \neq 0$. In this disc, $h(z)$ has a holomorphic logarithm, in other words, there exists a holomorphic function g such that $h(z) = e^{g(z)}$. Define $r(z) = e^{g(z)/k}$, and note that $r^k = h$. Therefore $f(z) = (zr(z))^k$, and $zr(z)$ has a simple zero at $z = 0$, thus $zr(z)$ is conformal. Therefore f can be viewed as the composition of $z \mapsto zr(z)$, which is conformal and $z \mapsto z^k$ which is k to one.

Let $z \in \mathcal{M}(X)$ be a fixed nonconstant function and let $deg((z)_0) = n$. Let h be any nonconstant meromorphic function on X. We will show that h satisfies a polynomial of degree n over $\mathbb{C}(z)$. Define X_0 to be the points of X at which dz does not have a zero or pole and let $\mathbb{C}_0 = z(X_0)$. Note that $X \setminus X_0$ is finite. Let $p \in X_0$, let U be a small neighborhood of p and let $\zeta_0 = z(p)$. Then there are n points of X_0, $p_1 = p, p_2, \ldots, p_n$, such that $z(p_i) = \zeta_0$. In addition there are n neighborhoods of those points $U_1 = U, U_2, \ldots, U_n$, such that $p_i \in U_i$ and $z |_{U_i}$ is a biholomorphism. Thus, locally, there are n inverse functions $f_i : z(U_i) \to U_i$ such that $q = f_i(\zeta)$ if and only if $\zeta = z(q)$ for all $q \in U_i$. Given a small neighborhood of ζ_0, the inverse functions are unique up to their numbering. Using $f_1, \ldots, f_n$, we construct n new functions, $g_1, \ldots, g_n : \mathbb{C}_0 \to \hat{\mathbb{C}}$ as follows:

$$g_1(\zeta) = -(h(f_1(\zeta)) + h(f_2(\zeta)) + \ldots h(f_n(\zeta))),$$

$$g_2(\zeta) = (-1)^2 \sum_{i<j} h(f_i(\zeta))h(f_j(\zeta)).$$

$$\vdots$$

$$g_n(\zeta) = (-1)^n h(f_1(\zeta))h(f_2(\zeta)) \cdots h(f_n(\zeta)).$$

For $i = 1, 2, \ldots, n$, the function g_i is the ith symmetric function on the set $\{h(f_1(\zeta)), h(f_2(\zeta)), \cdots, h(f_n(\zeta))\}$. Note that each g_i is invariant under a change in the numbering of the inverse functions f_i. Since h is meromorphic, and each f_i is locally holomorphic, we deduce that each g_i is locally meromorphic.

However, since at each ζ, $f_1, \ldots, f_n$ are unique up to their numbering, we see that each g_i is globally defined. Therefore g_i is a meromorphic function on C_0. On the other hand, at any point $p_0 \in X \setminus X_0$ where $ord_{p_0}(dz) = k - 1 > 0$, Proposition 12 yields that in a small neighborhood of p_0 the map given by z is k to one. Therefore, in a small neighborhood of $\zeta_0 = z(p_0)$, for each $\zeta \neq \zeta_0$ there will be k points, $q_1, q_2, \ldots, q_k$, near p_0 such that $z(q_i) = \zeta$ for each i. Since h is meromorphic, $h(q_i)$ will be near $h(p_0)$ for each i. This shows that each $g_i(\zeta)$ is continuous when $ord_{p_0}(dz) > 0$. A similar argument holds if $ord_{p_0}(z) < 0$, by considering local coordinates of the point at infinity of $\hat{C}$. Therefore each $g_i(\zeta)$ is a meromorphic function on $\hat{C}$. However, the meromorphic functions on $\hat{C}$ are the rational functions of ζ. Therefore, for each i, $g_i(\zeta) = r_i(\zeta)/s_i(\zeta)$ where r_i and s_i are polynomials with complex coefficients. But by composing this with z, we obtain $g_i(z) = r_i(z)/s_i(z) : X \to \hat{C}$ is a meromorphic function on X which can be expressed as a rational function in terms of the function z.

Now consider

$$(h - h(f_1(z)))(h - h(f_2(z))) \ldots (h - h(f_n(z))). \tag{1}$$

At each point of X_0, (1) equals zero. On the other hand, when expanded out, we obtain that 1 equals

$$h^n + g_1(z)h^{n-1} + \ldots + g_{n-1}h + g_n.$$

Since this vanishes on X_0 and h and each g_i are meromorphic functions on X, this must be identically zero on X. Thus h satisfies the minimal polynomial

$$G(H) = H^n + H^{n-1}r_1(z)/s_1(z) + \ldots + Hr_{n-1}(z)/s_{n-1}(z) + r_n(z)/s_n(z) = 0.$$

By multiplying $G(H)$ by the least common multiple of $s_1, s_2, \ldots, s_n$, we obtain that (z, h) satisfies a polynomial $F(Z, H) = 0$ of degree n in H.

We have proved the following theorem.

Theorem 13 *Let X be a compact Riemann surface of genus g. Let $z \in \mathcal{M}(X)$ be a nonconstant function and let $(z)_0$ have degree n. Let h be any nonconstant element of $\mathcal{M}(X)$. Then (z, h) satisfies a polynomial $F(Z, H) = 0$ of degree n in H. Thus $[\mathbb{C}(z, h) : \mathbb{C}(z)] \leq n$.*

Corollary 14 *Let X and z be as in the theorem. Then $[\mathcal{M}(X) : \mathbb{C}(z)] \leq n$.*

Proof. Assume first that $\mathcal{M}(X)$ is a finite field extension of $\mathbb{C}(z)$ and that $[\mathcal{M}(X) : \mathbb{C}(z)] > n$. From the theorem of the primitive element, since the characteristic of $\mathbb{C}(z)$ is 0, $\mathcal{M}(X)$ can be realized as a primitive field extension of $\mathbb{C}(z)$, thus $\mathcal{M}(X) = \mathbb{C}(z, g)$ for some $g \in \mathcal{M}(X)$. However, by Theorem 13, g satisfies a polynomial of degree n. This contradicts that $[\mathcal{M}(X) : \mathbb{C}(z)] > n$.

Now assume $\mathcal{M}(X)$ is an infinite field extension of $\mathbb{C}(z)$. In this case, the above argument has to be modified only slightly. Since $\mathcal{M}(X)$ is algebraic over

$\mathbb{C}(z)$, there are infinitely many field extensions K such that $\mathbb{C}(z) < K < \mathcal{M}(X)$ and $[K : \mathbb{C}(z)] > n$. We obtain a contradiction, as above, by applying the primitive element theorem to K.

Corollary 15 *Let X be a compact Riemann surface of genus g. Let $z \in \mathcal{M}(X)$ be a nonconstant function and let $deg((z)_0) = n$. Then $[\mathcal{M}(X) : \mathbb{C}(z)] = n$.*

Proof: Assume $[\mathcal{M}(X) : \mathbb{C}(z)] < n$. Let p be a point of X at which $\zeta = z(p) \neq \infty$ and $z^{-1}\{\zeta\}$ consists of n distinct points, say $p_1, \ldots, p_n$. From Corollary 11, there exists an $h \in \mathcal{M}(X)$ which takes on distinct values at these points, say $h(p_i) = a_i$. Then h satisfies a minimal polynomial $G(H)$ over $\mathbb{C}(z)$ of degree $k < n$. By clearing $G(H)$ of fractions, we obtain a polynomial $\hat{G}(Z, H)$ of degree k in H such that $\hat{G}(z, h)$ is the zero function. However $\hat{G}(\zeta, H) = 0$ for $H = a_1, \ldots, a_n$. This contradicts that the degree of $G(H)$ is k.

We have proved that the field of meromorphic functions of a compact Riemann surface yields an algebraic function field of one variable. In the next section we will prove the converse of this. Before we do, we conclude this section by once again emphasizing the correspondence between points on a Riemann surface and discrete valuation rings contained in its field of meromorphic functions.

Theorem 16 *Let X be a compact Riemann surface of genus g. Then each discrete valuation on $\mathcal{M}(X)$ is given by a point of X.*

Proof: Let v be a valuation on $\mathcal{M}(X)$ and let z be any element in $\mathcal{M}(X)$ such that $v(z) = 1$. Let $p_1, \ldots, p_j$ be the distinct zeros of z. Clearly, $z \notin \mathbb{C}$. From Proposition 6 we have $v(z^k r(z)/s(z)) = k$ if $r(z)$ and $s(z)$ are polynomials in z which are not divisible by z. Now let $f \in \mathcal{M}(X)$ be any nonconstant meromorphic function. Then f satisfies a minimal polynomial $f^n + t_1(z)f^{n-1} + \ldots + t_{n-1}(z)f + t_n(z) = 0$ for some rational functions $t_i \in \mathbb{C}(z)$. We will show that if $v(f) < 0$, then f has a pole at one of the p_i. If $v(f) < 0$, then Proposition 5 yields that $v(t_i) < 0$, for some i. However if $t_i(z) = z^k r_i(z)/s_i(z)$, where $r_i(z)$ and $s_i(z)$ are polynomials which are not divisible by z, then $v(t_i(z)) = k < 0$. Thus t_i has poles at $p_1, \ldots, p_j$. Recall from the arguments preceding Theorem 13 that t_i is a symmetric function of $f(p_1), \ldots, f(p_j)$, (counting multiplicities). If f were defined at each p_i, then t_i, being a symmetric function of $f(p_1), \ldots, f(p_j)$, would be defined. Therefore f must have a pole at one of $p_1, \ldots, p_j$. If $v(f) > 0$, then the above argument shows that $1/f$ must have a pole at one of $p_1, \ldots, p_j$, therefore f has a zero at one of $p_1, \ldots, p_j$. We conclude therefore, that if f has neither a zero nor pole at $p_1, \ldots, p_j$, then $v(f) = 0$.

Now pick $f_1, \ldots, f_j$ such that each f_i is defined at $p_1, \ldots, p_j$, f_i has a simple zero at p_i, and $f_i(p_{i'}) \neq 0$ if $i \neq i'$. We know this is possible from Corollary 10.

With z as above, let $e_i = ord_{p_i}(z)$. Then $g := z/(f_1^{e_1} \ldots f_j^{e_j})$ has no zeros or poles at $p_1, \ldots p_j$, thus $v(g) = 0$. This implies $1 = v(z) = \sum_{i=1}^{j} e_i v(f_i)$. Thus $v(f_i) = 1$ for precisely one i and $v(f_{i'}) = 0$ for $i \neq i'$. By renumbering, if necessary, we may assume $i = 1$.

We claim that $v = ord_{p_1}$. To see this let $h \in \mathcal{M}(X)$ be any function. We now define $e_i := ord_{p_i}(h)$ and $g := h/(f_1^{e_1} \ldots f_j^{e_j})$. Then g has no zero or pole at $p_1, \ldots, p_j$, thus $v(g) = 0$. Thus $v(h) = \sum_{i=1}^{j} e_i v(f_i) = e_1 v(f_1) = e_1$. Thus $v(h) = ord_{p_1}(h)$.

1.4   Algebraic Functions fields of one variable

We now reverse the above arguments. We begin with an algebraic function field of one variable K, and construct a Riemann surface X such that $\mathcal{M}(X) = K$. The key idea is that each point of a Riemann surface yields a discrete valuation ring and that all discrete valuation rings of $\mathcal{M}(X)$ arise in this manner. Roughly speaking, the discrete valuation rings become the points of the Riemann surface X. We follow the classic work [3].

Proposition 17 *Let K be an algebraic function field of one variable. Let R be a discrete valuation ring with maximal ideal I. Then R contains $\mathbb{C}$ and $R/I \cong \mathbb{C}$. If $r \in R$, then there exists $c \in \mathbb{C}$ such that $r - c \in I$.*

Proof: We proved in Proposition 3 that R contains $\mathbb{C}$. Since I is a maximal ideal of R, R/I is a field. On the other hand, R/I contains $\mathbb{C}$, so R/I is a field extension of $\mathbb{C}$. We will show that R/I is an algebraic extension of $\mathbb{C}$, thus $R/I \cong \mathbb{C}$.

Let t be a generator of I. Let g be any element of $R \setminus \mathbb{C}$. Since t is transcendental over $\mathbb{C}$, g is algebraic over $\mathbb{C}(t)$. Thus $g^n + r_{n-1}(t)g^{n-1} + \ldots + r_1(t)g + r_0(t) = 0$ for some rational functions r_i. Let $r_i(t) = t^{k_i}p_i(t)/q_i(t)$ where p_i and q_i are relatively prime polynomials not divisible by t. Let $k = min\{k_0, \ldots, k_n\}$. If $k > 0$, then for each i, $r_i \in I$, thus $g + I$ is algebraic over $\mathbb{C}$. If $k \leq 0$, then $t^{-k}(g^n + r_{n-1}(t)g^{n-1} + \ldots + r_1(t)g + r_0(t)) = 0$, however $t^{-k}r_i(t) \in R \setminus I$ for some i, thus $g + I$ is algebraic over $\mathbb{C}$. In either case, since $\mathbb{C}$ is algebraically closed, $g + I \in \mathbb{C}$. Thus $R/I \cong \mathbb{C}$. Given any $r \in R$, there exists a $c \in \mathbb{C}$ such that $r + I = c + I$, therefore $r - c \in I$.

Let R be a discrete valuation ring, let I be the unique maximal ideal of R, and let $g \in K$. We will use the above proposition to obtain a map $K \to \hat{\mathbb{C}}$ as follows. If $g \in R$, we map $g \mapsto g + I \in \mathbb{C}$. If $g \notin R$, we map $g \mapsto \infty$.

We now establish some notation used throughout this section. Let K be an algebraic function field of one variable. Each discrete valuation ring of K has a unique maximal ideal; these maximal ideals are called places of K and the set of places will be denoted by $\mathcal{S}(K)$. Below we will give $\mathcal{S}(K)$ a topology which will make it a Riemann surface. If ρ is a place, we will denote the discrete valuation ring associated to it by R_ρ. If $x \in K$ and $\rho \in \mathcal{S}(K)$, we define $\pi_x : \mathcal{S}(K) \to \hat{\mathbb{C}}$, by $\pi_x(\rho) = x + \rho \in R/\rho \cong \mathbb{C}$, if $x \in R_\rho$, or $\pi_x(\rho) = \infty$ if $x \notin R_\rho$. Throughout this section, π_x will always refer to this map. If ρ is held fixed and x is allowed to vary, π_x induces a map from K to $\hat{\mathbb{C}}$.

We now make $\mathcal{S}(K)$ into a Riemann surface by choosing a topology on $\mathcal{S}(K)$ to ensure that $\pi_x : \mathcal{S}(K) \to \hat{\mathbb{C}}$ is continuous for each $x \in K$. We define the topology on $\mathcal{S}(K)$ by defining a subbasis for the topology to be all sets of the form $\pi_x^{-1}(U)$, where $x \in K$ and U is an open subset of $\hat{\mathbb{C}}$. An open set of $\mathcal{S}(K)$ is the union of sets of the following form:

$$\pi_{x_1}^{-1}(U_1) \bigcap \pi_{x_2}^{-1}(U_2) \bigcap \cdots \bigcap \pi_{x_n}^{-1}(U_n)$$

where each U_i is open in $\hat{\mathbb{C}}$ and $x_1, x_2, \ldots, x_n$ are elements of K. In order to prove that S is compact however, we choose a different way of describing this topology.

Let $\hat{\mathbb{C}}_x$ be a copy of $\hat{\mathbb{C}}$. Let P denote the product space

$$\prod_{x \in K} \hat{\mathbb{C}}_x.$$

We obtain a map $\Psi : \mathcal{S}(K) \to P$ defined as follows. For each ρ, we define the x coordinate of $\Psi(\rho)$ to be the element $\pi_x(\rho)$. Let S' denote the image of Ψ. Observe that Ψ is one to one, since for distinct places ρ_1 and ρ_2 there exists an $x \in K$ such that $\pi_x(\rho_1) \neq \pi_x(\rho_2)$. Every open set of P is of the form

$$\prod_{x \in K} U_x,$$

where U_x is open in $\hat{\mathbb{C}}_x$ and $U_x = \hat{\mathbb{C}}_x$ for all but a finite number of indices. Thus, if we identify $\mathcal{S}(K)$ and S', we see that the topology defined above on $\mathcal{S}(K)$ is precisely the subspace topology on S' induced by P. To prove that $\mathcal{S}(K)$ is compact, it will be easier to prove that S' is compact.

First note that since P is a Hausdorf space, S' is a Hausdorf space. In addition, since P is the product of compact spaces, P itself is compact. To show that $\mathcal{S}(K)$ is compact, we will prove that S' is closed in P.

Proposition 18 *Let $\rho \in \mathcal{S}(K)$ and let x and y be elements of K. Let $\Psi(\rho) = \alpha = \{\alpha_x\}_{x \in K}$ be an element of S'. Assume α_x and α_y are not infinity. Then $\alpha_{x-y} = \alpha_x - \alpha_y$, $\alpha_{xy} = \alpha_x \alpha_y$, and if $a_x \neq 0$, then $\alpha_{1/x} = 1/\alpha_x$. In addition, if $\alpha_x = \infty$, then $\alpha_{1/x} = 0$ and if $\alpha_x = 0$, then $\alpha_{1/x} = \infty$.*

Proof: Assume $\alpha_x \neq \infty$ and $\alpha_y \neq \infty$. Since $\alpha = \Psi(\rho)$, $\alpha_x = x + \rho$ and $\alpha_y = y + \rho$. Since R_ρ is a ring and ρ is an ideal of R_ρ, we obtain that $\alpha_{x-y} = x - y + \rho = x + \rho - y + \rho = \alpha_x - \alpha_y$. Analogously, $\alpha_{xy} = xy + \rho = (x+\rho)(y+\rho) =$

$\alpha_x \alpha_y$. Also, if $\alpha_x \neq 0$, then $1/x - 1/\alpha_x = (\alpha_x - x)/(\alpha_x x)$. Since $\alpha_x x \notin \rho$, and $\alpha_x - x \in \rho$, we deduce that $1/x - 1/\alpha_x \in \rho$. Thus $\alpha_{1/x} = 1/\alpha_x$. If $\alpha_x = \infty$, then $x \notin R_\rho$. Thus, from Proposition 2, $1/x \in \rho$. Thus $\alpha_{1/x}(\rho) = 0$. If $\alpha_x = 0$, then $1/x \notin R_\rho$, thus $\alpha_{1/x} = \infty$.

We now prove that S' is closed in P. Assume that $\alpha = \{\alpha_x\}_{x \in K} \in P$ is a limit point of S'. We will show that $\alpha \in S'$. To do this, we need to show that there is a place ρ such that $\Psi(\rho) = \alpha$, in other words, that $\pi_x(\rho) = \alpha_x$ for each $x \in K$. Let α^k, with $k = 1, 2, \ldots \infty$, be a sequence of points of S' which converge to α. Then there is a sequence of places $\rho_1, \rho_2, \ldots$ such that $\Psi(\rho_k) = \alpha^k$. We denote the x coordinate of α^k by α_x^k and note that it equals $\pi_x(\rho_k)$.

We define a set $R \subset K$. Let $x \in R$ if and only if $\alpha_x \neq \infty$. We will show that R is a discrete valuation ring. From Proposition 2, it is sufficient to show that R is a ring which contains $\mathbb{C}$ and that for each $x \in K$, if $x \notin R$, then $x^{-1} \in R$.

To show that R contains $\mathbb{C}$, let $c \in \mathbb{C}$. Then for each sequence α^k, which converges to α, we have that $\alpha_c^k = \pi_c(\rho_k) = c$. Thus $\alpha_c = \lim_{k \to \infty} \alpha_c^k = c$. Thus $c \in R$.

We now show that R is a ring. Let x and y be elements of R, thus $\alpha_x \neq \infty$ and $\alpha_y \neq \infty$. We must show that $x - y$ and $xy \in R$. Since $\Psi(\rho_k) = \alpha^k$ is a sequence of points of S' converging to α, for large enough k, α_x^k and α_y^k will not be infinity. Thus Proposition 18 yields that $\alpha_{x-y}^k = \alpha_x^k - \alpha_y^k$ and $\alpha_{xy}^k = \alpha_x^k \alpha_y^k$ for all large k. Thus $\alpha_x - \alpha_y = \lim_{k \to \infty}(\alpha_x^k - \alpha_y^k) = \lim_{k \to \infty}(\alpha_{x-y}^k) = \alpha_{x-y}$. Thus $\alpha_{x-y} \neq \infty$. A similar calculation holds for α_{xy}. Thus R is a ring.

We now show that R is a discrete valuation ring. Assume $z \notin R$. We will show $z^{-1} \in R$. Since $z \notin R$, $\alpha_z = \infty$, thus for large k, $\pi_z(\rho_k)$ is large or ∞. Thus Proposition 18 yields that $\alpha_{1/z} = \lim_{k \to \infty} \alpha_{1/z}^k = 0$. Thus $1/z \in \rho \subset R$. Thus R is a discrete valuation ring. Let ρ be the place corresponding to R. We must show that $\Psi(\rho) = \alpha$. However, if $\alpha_x \neq 0$, then $\alpha_{1/x} \neq 0$, thus x is a unit of R. Therefore $x \in \rho$ if and only if $\alpha_x = 0$. Assume $\alpha_x \neq \infty$. Since $\alpha_x \in \mathbb{C}$, $\alpha_{(x-\alpha_x)} = \lim_{k \to \infty} \alpha_{x-\alpha_x}^k = \lim_{k \to \infty}(\alpha_x^k - \alpha_x) = 0$, thus $x - \alpha_x \in \rho$. Thus $\pi_x(\rho) = \alpha_x$ for all $x \in R$, thus $\alpha = \Psi(\rho)$.

To give $\mathcal{S}(K)$ the structure of a Riemann surface we will use the following theorems. The proofs may be found in [3].

Theorem 19 *Let K be an algebraic function field in one variable over $\mathbb{C}$. Let $x \in K \setminus \mathbb{C}$ and let $n = [K : \mathbb{C}(x)]$. Then there are a finite number of places $\rho_1, \ldots, \rho_k$ which contain x. In addition, if $\mathrm{ord}_{\rho_i}(x) = e_i$, then $\sum e_i = n$.*

Theorem 20 *Let $\rho_1, \ldots, \rho_n$ be distinct places and let $\alpha_1, \ldots, \alpha_n$ be distinct complex numbers. Then there exists $y \in K$ such that $\mathrm{ord}_{\rho_i}(y - \alpha_i) = 1$ for each $i = 1, \ldots, n$.*

Theorem 21 *Let $x, y \in K \setminus \mathbb{C}$. Let $F(x, y)$ be a minimum polynomial for y over $\mathbb{C}(x)$. Assume $F(\alpha, \beta) = 0$ for the complex numbers α and β. Then there is a place of K which contains $x - \alpha$ and $y - \beta$.*

We now show that each point $\rho \in \mathcal{S}(K)$ has a neighborhood homeomorphic with a neighborhood of 0 in the complex plane. In fact, we will show that if $x \in K$ with $ord_\rho(x) = 1$, then there is a neighborhood U of ρ such that $x : U \to \mathbb{C}$ is a homeomorphism of U with its image.

Let $\rho \in S$ and let $x \in K$ with $ord_\rho(x) = 1$. Let $\rho_1 = \rho, \rho_2, \ldots, \rho_k$ be the distinct places of K which contain x, and let $e_i = ord_{\rho_i}(x)$. We know from Theorem 19 that $n := [K : \mathbb{C}(x)] = e_1 + \ldots + e_k$. Let $\beta_1, \ldots, \beta_k$ be distinct complex numbers. From Theorem 20 we may choose $y \in K$ such that $ord_{\rho_i}(y - \beta_i) = 1$ for $i = 1, \ldots, k$. Since K is an algebraic function field in one variable, y satisfies an irreducible polynomial $F(X, Y)$ over $\mathbb{C}(x)$. Since $F(x, y) = 0$, we see that $0 + \rho_i = F(x, y) + \rho_i = F(0, \beta_i)$, thus β_i is a root of $F(0, Y)$. Let m_i be the multiplicity of the root β_i in $F(0, Y)$. We will prove that $m_i = e_i$. Note that $ord_{\rho_i}(F(0, y)) = ord_{\rho_i}(F(0, \beta_i)) = m_i$. Note that $F(X, Y) - F(0, Y)$ is divisible by X. Since $F(x, y)$ is identically 0, we have that $ord_{\rho_i}(F(x, y) - F(0, y)) = ord_{\rho_i}(F(0, y)) = m_i$. On the other hand, since X divides $F(X, Y) - F(0, Y)$ we have that $ord_{\rho_i}(F(x, y) - F(0, y)) \geq ord_{\rho_i}(x) = e_i$. However, the sum of the m_i is less than or equal to n, and $e_1 + \ldots + e_k = n$. This forces $m_i = e_i$ for each i. In addition, this says that $F(X, Y)$ has degree n, thus $K = \mathbb{C}(x, y)$.

We use the above to determine a neighborhood of ρ_1 which is small enough to guarantee that it maps topologically onto a neighborhood of the complex plane. For each i, let U_i be a neighborhood of β_i chosen small enough so that U_i and U_j are disjoint if $i \neq j$. Then since the roots of an algebraic equation are continuous, there exists a real number $r > 0$ with the following property: For each α with $0 <| \alpha |\leq r$, the equation $F(\alpha, Y)$ has exactly e_i roots, $\beta_{1,i}, \ldots \beta_{e_i,i}$, in U_i. From Theorem 21, there exist places $\rho_{i,j}$ which correspond to each pair $(\alpha, \beta_{i,j})$. Since $[K : \mathbb{C}(x)] = n$, the only places which contain $x - \alpha$ are the places $\rho_{i,j}$. Specifically, since $e_1 = 1$, we have that for each α such that $| \alpha |\leq r$ there is a unique $\beta \in U_1$. Furthermore, each such (α, β) corresponds to a unique place. Let U be this set of places. Thus U maps one to one and onto the set $| \alpha |\leq r$ of $\mathbb{C}$. This map is continuous. Since U is closed in S, it is compact, therefore the map is a homeomorphism. Therefore, each point of S has a neighborhood which maps topologically onto a neighborhood of 0 of the complex plane.

We now show that each element of K can be considered as a meromorphic function of $\mathcal{S}(K)$. Let $x \in K$ and let ρ_0 be a place such that $ord_{\rho_0}(x) = 1$. Then x yields a map $\gamma = x(\rho) : U \to V \subset \mathbb{C}$, from some neighborhood of ρ_0 onto V, a neighborhood of $\gamma = 0$ in the complex plane. Let $y \in K$. Then y induces a map $\gamma = y(\rho)$, from U to $\hat{\mathbb{C}}$ by $y(\rho) = y + \rho \in \mathbb{C}$, if $y \in R_\rho$ and

$y(\rho) = \infty$ if $y \notin R_\rho$. Thus we have an induced function $\hat{y} : V \to \hat{\mathbb{C}}$ defined by $\hat{y}(\gamma) = y \circ x^{-1}(\gamma)$. We will prove that this map is a meromorphic function on V. Let $F(X, Y)$ be an irreducible polynomial such that $F(x, y) = 0$ in K. Then $dF/dY(x, y)$ has a finite number of zeros. Since y has a finite number of poles, there are a finite number of γ such that $\hat{y}(\gamma) = \infty$. By removing these points, the implicit function theorem yields that $\hat{y}$ is a holomorphic function of γ in this restricted domain. However, as we have seen before in similar arguments, $\hat{y}$ is continuous on all of V, therefore it is a meromorphic function of γ.

Proposition 22 *Let $\rho \in \mathcal{S}(K)$, let x and y be elements of K, and let $ord_\rho(x) = 1$. Let U be a neighborhood of ρ and let x map U topologically onto V, a neighborhood of $\gamma = 0$ in the complex plane, so that $0 = x(\rho)$. Note that y induces a meromorphic function $\hat{y} = y \circ x^{-1}$ on V. Then $ord_\rho(y)$ equals the order of $\hat{y}$ when considered as a meromorphic function on V.*

Proof: Assume $y = x^k h$, where h is a unit in R_ρ. Note that $h + \rho \neq 0 + \rho = 0$. Therefore, $\hat{y}(\gamma) = y \circ x^{-1}(\gamma) = \gamma^k h(x^{-1}(\gamma))$, and $h((x^{-1}(0)) = h(\rho) \neq 0$. Thus, when considered as a meromorphic function of γ, $\hat{y}$ has order k.

Assume $\rho \in \mathcal{S}(K)$, and x and y are two elements of K such that $ord_\rho(x) = ord_\rho(y) = 1$. Then we may consider x as a local parameter at ρ, and y as a meromorphic function on $\mathcal{S}(K)$. From Proposition 22 , $\hat{y}$, when considered as a meromorphic function near ρ, has order one. Therefore $\hat{y}$ is a conformal mapping near ρ. However, since $ord_\rho(y) = 1$, y could have been chosen as a local parameter at ρ. This shows that changes of coordinates on $\mathcal{S}(K)$ are conformal, thus $\mathcal{S}(K)$ is a Riemann surface.

We have shown that $\mathcal{S}(K)$ is a Riemann surface and that each element of K is a meromorphic function on $\mathcal{S}(K)$. Thus, if $\mathcal{M}(\mathcal{S}(K))$ denotes the field of meromorphic functions on $\mathcal{S}(K)$, then $K \subset \mathcal{M}(\mathcal{S}(K))$. We now show that equality must hold. Let x be a nonconstant element of K. Considering x as an element of $\mathcal{M}(\mathcal{S}(K))$, let $p_1, \ldots, p_k$ be the distinct zeros of x and let $e_1, \ldots, e_k$ be the respective orders of x, considered as a meromorphic function, at each p_i. Define $n = e_1 + \ldots + e_k$. Then $[\mathcal{M}(\mathcal{S}(K)) : \mathbb{C}(x)] = n$. However, since each point of a Riemann surface corresponds to a discrete valuation ring, for each point p_i there corresponds a place ρ_i in $\mathcal{S}(K)$. From Proposition 22, $ord_{\rho_i}(x) = e_i$. Thus from Theorem 19 we obtain that $[K : \mathbb{C}(x)] = e_1 + \ldots + e_k = n$. Thus $K = \mathcal{M}(\mathcal{S}(K))$.

We have shown that if K is an algebraic function field of one variable, then there is a Riemann surface $\mathcal{S}(K)$ such that $K = \mathcal{M}(\mathcal{S}(K))$. On the other hand, we showed that if X is a Riemann surface, then $\mathcal{M}(X)$ is an algebraic function field of one variable. We now show that $\mathcal{S}(\mathcal{M}(X)) = X$. This is very easily done. The points of $\mathcal{S}(\mathcal{M}(X))$ are the places of $\mathcal{M}(X)$. However, we know that the places of $\mathcal{M}(X)$ are in one to one correspondence with the points of X; this gives us a map $\phi : \mathcal{S}(\mathcal{M}(X)) \to X$. We must show that ϕ is conformal, in other words, that ϕ is conformal when expressed in terms of local coordinates of $\mathcal{S}(\mathcal{M}(X))$ and X. Assume $\phi(\rho) = p$. As a local parameter at p, we may choose a function $z \in M(X)$ such that $ord_p(z) = 1$. On the other hand, if $ord_p(z) = 1$, then $ord_\rho(z) = 1$. Thus z is a local parameter at ρ. Thus ϕ, when expressed in terms of z, is the identity map. Thus $\mathcal{S}(\mathcal{M}(X)) = X$.

We make the following observations.

Theorem 23 *Let $\sigma : X \to Y$ be a nonconstant holomorphic function of Riemann surfaces. Then there is an induced field embedding $\sigma^* : \mathcal{M}(Y) \to \mathcal{M}(X)$, given by $g \mapsto g \circ \sigma$. If $\sigma_1 : X \to Y$ is a different holomorphic function, then $\sigma^* \neq \sigma_1^*$.*

Proof: Since $g \in \mathcal{M}(Y)$ is meromorphic and σ is holomorphic, clearly $g \circ \sigma \in \mathcal{M}(X)$. On the other hand σ^* is clearly an embedding of fields since $(f + g) \circ \sigma = f \circ \sigma + g \circ \sigma$ and $(fg) \circ \sigma = (f \circ \sigma)(g \circ \sigma)$. If $\sigma \neq \sigma_1$, then there is a point $p \in X$, such that $q = \sigma(p) \neq \sigma_1(p) = q_1$. Let $g \in \mathcal{M}(Y)$ be any function with $g(q) \neq g(q_1)$. Then $\sigma^*(g)(p) = g \circ \sigma(p) = g(q)$, while $\sigma_1^*(g)(p) = g \circ \sigma_1(p) = g(q_1)$. Thus $\sigma_1^* \neq \sigma^*$.

Theorem 24 *Let $\tau : \mathcal{M}(Y) \to \mathcal{M}(X)$ be an embedding of fields. Then there exists a holomorphic function $\sigma : X \to Y$ such that $\sigma^* = \tau$.*

Proof: We may assume that $\mathcal{M}(Y) \subset \mathcal{M}(X) = K$. We will identify the points of the Riemann surfaces X and Y with their places in $\mathcal{M}(X)$ and $\mathcal{M}(Y)$. If $g \in \mathcal{M}(Y)$, then for each place $\rho \in Y$, $g(\rho) = g + \rho$ if $g \in R_\rho$ and $g(\rho) = \infty$ if $g \notin R_\rho$. However, since $g \in \mathcal{M}(X)$, g also yields a meromorphic function on X. Thus for each place $\rho' \in X$, $g(\rho') = g + \rho'$ if $g \in R_{\rho'}$, and $g(\rho') = \infty$ if $g \notin R_{\rho'}$. When we consider g as an element of $\mathcal{M}(X)$, we will denote it by $\hat{g}$. Thus $\tau(g) = \hat{g}$. For each place of $\mathcal{M}(Y)$, there are a finite number of places of $\mathcal{M}(X)$ which contain it. In addition, given a place in $\mathcal{M}(X)$, there is a unique place of $\mathcal{M}(Y)$ which is contained in it. Let $\rho \in Y$, and let $\rho_1, \ldots, \rho_k$ be the places of $\mathcal{M}(X)$ which contain it. Then we construct a map $\sigma : X \to Y$ by $\sigma(\rho_i) = \rho$. We must show that $\tau(g) = \hat{g} = g \circ \sigma$ and that σ is holomorphic.

Assume that $g \in R_{\rho_i}$, thus $\hat{g}(\rho_i) = g + \rho_i$. On the other hand, $g \circ \sigma(\rho_i) = g(\rho) = g + \rho$. If g is defined at ρ, $g + \rho = \alpha \in \mathbb{C}$. Then $g - \alpha \in \rho$, thus $g - \alpha \in \rho_i$, since $\rho \subset \rho_i$. Thus $\alpha = g \circ \sigma(\rho_i) = \hat{g}(\rho_i)$. Thus, in this case, $\hat{g} = g \circ \sigma$. If g is not defined at ρ, then $g \notin R_\rho$, thus $1/g \in \rho$, so $1/g \in \rho_i$, which contradicts that $g \in R_{\rho_i}$. Thus if $\hat{g}(\rho_i) \neq \infty$, then $\hat{g}(\rho_i) = g \circ \sigma(\rho_i)$. Now assume that $\hat{g}(\rho_i) = \infty$. We will show that $g(\rho) = \infty$. If not, then $g(\rho) = \alpha \neq \infty$, so $g - \alpha \in \rho$, so $g - \alpha \in \rho_i$ which contradicts that $g \notin R_{\rho_i}$. Thus $g(\rho) = \infty$, thus $\hat{g}(\rho_i) = g \circ \sigma(\rho_i)$. Thus, $\hat{g} = g \circ \sigma$.

We must now show that σ is holomorphic, in other words, σ is a holomorphic function when expressed in terms of local coordinates for X and Y. Let $\rho_i \in X$, and let $\sigma(\rho_i) = \rho$. Let $g \in M(Y)$ be a local parameter at ρ. We must show that $g \circ \sigma$ is holomorphic at ρ_i. However $g \circ \sigma = \hat{g} = \tau(g)$ which is a meromorphic function of X by assumption. Since $g \circ \sigma$ is defined at ρ, it is holomorphic. Thus σ is holomorphic.

The previous two theorems show that the assignment $\sigma \mapsto \sigma^*$ is both one to one and onto. We have proven the following theorem.

Theorem 25 *The category of compact Riemann surfaces is functorially equivalent to the category of algebraic function fields of one variable over $\mathbb{C}$. For each Riemann surface X, let $\mathcal{M}(X)$ denote its field of meromorphic functions. Then $\mathcal{M}(X)$ is an algebraic function field of one variable over $\mathbb{C}$. To each algebraic function field of one variable K, let $\mathcal{S}(K)$ be the Riemann surface defined in this section. Then $\mathcal{S}(\mathcal{M}(X)) = X$ and $\mathcal{M}(\mathcal{S}(K)) = K$. If X and Y are Riemann surfaces and if $\sigma : X \to Y$ is a holomorphic map, then there is a unique field embedding $\sigma^* : \mathcal{M}(Y) \to \mathcal{M}(X)$, such that $\sigma^*(g) = g \circ \sigma$ for each $g \in \mathcal{M}(Y)$. Every field embedding of $\mathcal{M}(Y)$ into $\mathcal{M}(X)$ arises in this way.*

We prove one final theorem which will be useful in the example considered in Section II.

Definition 3 *Let $F(X, Y)$ be a polynomial of two variables with complex coefficients, and assume that $F(\alpha, \beta) = 0$. Then $F(X, Y)$ can be expressed as a polynomial in $X - \alpha$ and $Y - \beta$. We define the tangent line of $F(X, Y)$ at (α, β) to be the linear term of $F(X, Y)$ when expressed in terms of $X - \alpha$ and $Y - \beta$. Note that the linear term is not zero if and only if $F(X, Y)$ is nonsingular at (α, β).*

Theorem 26 *Let $K = \mathbb{C}(x, y)$ be an algebraic function field of one variable, and let $F(X, Y)$ be an irreducible polynomial with complex coefficients such that $F(x, y) = 0$. Let V be the set of points in $\mathbb{C}^2$ which satisfy $F(X, Y) = 0$. Let $P = (\alpha, \beta)$ be a point of V at which F is nonsingular, meaning $dF/dX(\alpha, \beta)$ and $dF/dY(\alpha, \beta)$ are not both equal to zero. Let $A(X - \alpha) + B(Y - \beta)$ be the tangent line of $F(X, Y)$ at P and let $O_P = \{G(x, y)/H(x, y) \mid H(\alpha, \beta) \neq 0\}$. Then O_P is a discrete valuation ring. In addition, if I_P is the maximal ideal of O_P, then I_P is generated by any linear combination of $(x - \alpha)$ and $(y - \beta)$ which is not a multiple of $A(x - \alpha) + B(y - \beta)$.*

Proof: Since F is nonsingular, it has a linear term. Assume the linear term is $A(X - \alpha) + B(Y - \beta)$. By a change of coordinates, we may assume that $P = (0, 0)$ and that the linear term of F is cY, for some complex number c. Let I be the ideal of O_P generated by (x, y). It is clear that I is maximal, since if $G(x, y)/H(x, y) \in O_P$ and $G(X, Y)$ has a constant term, then $G(0, 0) \neq 0$, thus $G(x, y)/H(x, y)$ is invertible.

Since cY is the tangent line, $F(x, y)$ can be expressed in the form $F(x, y) = yf(x, y) + x^k h(x)$ for some polynomials $f(X, Y)$ and $h(X)$ such that $f(X, Y)$ and $h(X)$ have nonzero constant terms. In particular since $F(x, y) = 0$, we have that $y = -x^k h(x)/f(x,y)$. Since $f(0,0) \neq 0$, we have that $h(x)/f(x,y) \in O_P$, thus $y \in (x)$. Thus I_P is principally generated. Therefore O_P is a discrete valuation ring.

Each nonsingular point of a plane curve corresponds to a point of a Riemann surface. Theorem 26 gives a criterion to determine which functions generate the corresponding place. At singular points, there are a finite number of points of a Riemann surface "lying over" it. We will employ this in analyzing a specific Riemann surface below.

2 Example

Let $\mathcal{X}$ be the Riemann surface whose field of meromorphic functions satisfies the irreducible equation

$$F(x,y) := y^9 + Ax^2y^3 + Bx^7 + Cx^2y = 0, \tag{2}$$

where A, B, and C are complex numbers which satisfy certain mild restrictions discussed below. One common way to analyze $\mathcal{X}$ is to consider the locus V of F in $\mathbb{C}^2$, explicitly determine places where the map $(x,y) \mapsto x$ is branched, determine cuts, and glue the cuts together to construct a compact Riemann surface. In this section we take a different approach. We consider the Riemann surface $\mathcal{X}$ as already existing, and that F is one of many equations which define it. At a point of V where F is nonsingular, Theorem 26 yields that there is a corresponding point of $\mathcal{X}$. At a point of V where F is singular, V does not effectively represent $\mathcal{X}$. At such a point of V, there may be one or several points of $\mathcal{X}$ corresponding to it. However, one can find several defining equations for $\mathcal{X}$, no two of which have the same singular points. In this way, we can determine a good, local representation at each point of $\mathcal{X}$.

The only restrictions we place on A, B, and C in F is that B and C are nonzero and A is chosen so that $F(X,Y)$ is nonsingular at all points $(\alpha,\beta) \neq (0,0)$. We now show that all but a finite number of values of A satisfy this condition. First note that F is singular at $(0,0)$. Let (α,β) be any other point of V. If (α,β) is singular, then $F(\alpha,\beta) = F_X(\alpha,\beta) = F_Y(\alpha,\beta) = 0$, thus

$$\alpha F_X(\alpha,\beta) = 2A\alpha^2\beta^3 + 7B\alpha^7 + 2C\alpha^2\beta \tag{3}$$

and

$$\beta F_Y(\alpha,\beta) = 9\beta^9 + 3A\alpha^2\beta^3 + C\alpha^2\beta, \tag{4}$$

so

$$6F - 3\alpha F_X = 6\beta^9 - 15B\alpha^7, \quad 3F - \beta F_Y = -6\beta^9 + 3B\alpha^7 + 2C\alpha^2\beta. \tag{5}$$

Thus $0 = 2\alpha^2(-6B\alpha^5 + C\beta)$, so $\beta = 6B\alpha^5/C$. Substituting this value into the first equation of (5), we note that there are only a finite number of values of

α which satisfy it. Now A can be chosen so that none of these values of α satisfies (2). Thus F is nonsingular at all points $(\alpha, \beta) \neq (0,0)$.

In what follows, if $G(X, Y)$ is a polynomial of two variables, we let G_x and G_y denote $G_X(x,y)$ and $G_Y(x,y)$ respectively.

We will determine the genus of $\mathcal{X}$, a basis for the holomorphic differentials of $\mathcal{X}$, and the full automorphism group of $\mathcal{X}$. To accomplish this, we recall some basic facts concerning the Weierstrass points of a Riemann surface. Let P be a point of an arbitrary Riemann surface Z of genus $g \geq 2$. Employing the notation of Section 1.2, $l(0) = 1$, and $l((2g-1)P) = g$. Considering the set of divisors $\{0, P, 2P, \ldots, kP, \ldots, (2g-1)P\}$, there are g positive integers $1 = k_1 < k_2 < \ldots < k_g < 2g$ with the property that, for each i with $0 \leq i \leq g$, there is no meromorphic function of Z such that $(f)_\infty = k_i P$. We call the sequence $(k_1, \ldots, k_g)$ the gap sequence at P. For all but a finite number of points of Z, the gap sequence is $(1, 2, \ldots, g)$, however there always exist points, called Weierstrass points, with a different gap sequence. Under an automorphism of Z, Weierstrass points get mapped to Weierstrass points with an identical gap sequence. The following proposition will allow us to compute gaps at various points of $\mathcal{X}$.

Proposition 27 *Let Z be a Riemann surface of genus $g \geq 2$, and let $W = n_1 P_1 + \ldots + n_r P_r$ be a canonical divisor of Z. Assume $P \notin \{P_1, \ldots, P_r\}$. Then k is a gap for P if and only if there is a meromorphic function in $L(W)$ which has a zero of order exactly $k - 1$ at P.*

Proof: It is clear that k is a gap at P if and only if $l(kP) = l((k-1)P)$. However the Riemann Roch theorem yields that $l(W - kP) = 2g - 2 - k + 1 - g + l(kP)$ and $l(W - (k-1)P) = 2g - 2 - (k-1) + 1 - g + l((k-1)P)$. Therefore k is a gap if and only if $l(W - kP) - 2g + 2 + k - 1 + g = l(W - (k-1)P) - 2g + 2 + (k-1) - 1 + g$. Therefore k is a gap at P if an only if $l(W - kP) + 1 = l(W - (k-1)P)$, in other words, if and only if there is a meromorphic function in $L(W)$ which has a zero of order exactly $(k-1)$ at P.

Note that Proposition 27 can be modified slightly to yield the following. With W as in Proposition 27, if $P = P_i$, then $k > n_i$ is a gap at P if and only if $l(W - kP) + 1 = l(W - (k-1)P)$, in other words, if and only if there is an $f \in L(W)$ which has a zero of order exactly $k - 1 - k_i$ at P.

We now examine the singular point $(\alpha, \beta) = (0,0)$. There must be at least one point $p \in \mathcal{X}$ where the functions x and y satisfy $x(p) = y(p) = 0$. Any point p with this property is said to "lie over" $(0,0)$. We determine how many points have this property and what the orders of x and y are at each point. In doing this we will also determine new defining equations for $\mathcal{X}$ so that each point p which lies over $(0,0)$ will be nonsingular.

Considering F, we note that, in general, there are 9 points of $\mathcal{X}$ where x takes on a specific value. Thus $deg((x)_0) = 9$ and $[C(x,y) : C(x)] = 9$. Similarly, $deg((y)_0) = 7$. Thus there must be a $p \in \mathcal{X}$, where $x(p) = 0$ and

$ord_p(x) \geq ord_p(y) > 0$. We re-examine (2) at p and emphasize that $F(x, y)$ is the zero function on $\mathcal{X}$. Note that $ord_p(Cx^2y)$ must be strictly less than $ord_p(Ax^2y^3)$ and $ord_p(Bx^7)$. Since the defining equation is the zero function on $\mathcal{X}$, Proposition 4 yields that $ord_p(Cx^2y) = ord_p(y^9)$. Therefore $2ord_p(x) + ord_p(y) = 9ord_p(y)$, thus $4ord_p(y) = ord_p(x)$. Recall that $deg((x)_0) = 9$, thus $ord_p(y) = 1$ or 2. We make a change of variables to examine p. Define $s = x/y^4$. Note that $\mathbb{C}(x, y) = \mathbb{C}(s, y)$, thus $\mathcal{X}$ is the Riemann surface associated to $\mathbb{C}(s, y)$. Substituting $sy^4 = x$ into (2) yields:

$$y^9(1 + As^2y^2 + Bs^7y^{19} + Cs^2) = 0. \tag{6}$$

Since y^9 is not the zero function, the second factor of (6) is the zero function. At p, both $y(p) = 0$ and $(1+As^2y^2+Bs^7y^{19}+Cs^2)(p) = 0$. Thus $s(p)^2 = -1/C$, so s can take on two values when y is 0. Thus there must be two points on $\mathcal{X}$ where y is 0 and $4ord_p(y) = ord_p(x)$. Let P denote the point where $y(P) = 0$ and $s(P) = \sqrt{-1/C}$ and let Q be the point where $s(Q) = -\sqrt{-1/C}$. Since $s = x/y^4$ we have that $ord_P(x) = 4$, $ord_P(y) = 1$ and $x/y^4(P) = \sqrt{-1/C}$ while, $ord_Q(x) = 4$, $ord_Q(y) = 1$ and $x/y^4(Q) = -\sqrt{-1/C}$.

Since $deg((x)_0) = 9$, this immediately yields that there is another point $R \in \mathcal{X}$ where x has a zero of smaller order than y does. Although we can immediately deduce that $ord_R(x) = 1$ and $ord_R(y) = 5$, we reaffirm this by examining F. From F we conclude, in conjunction with Proposition 4, that $ord_R(Cx^2y)$ is less than $ord_R(y^9)$ and $ord_R(Ax^2y^3)$. Thus $ord_R(x^7) = ord_R(x^2y)$. Thus $7ord_R(x) = 2ord_R(x)+ord_R(y)$, so $ord_R(x) = 1$ and $ord_R(y) = 5$. Define $t = y/x^5$, and note that $\mathbb{C}(x, t) = \mathbb{C}(x, y)$, thus $\mathcal{X}$ is the Riemann surface associated to $\mathbb{C}(x, t)$. By noting that $tx^5 = y$, we may rewrite (2) to obtain

$$x^7(t^9y^{38} + At^3x^{10} + B + Ct) = 0. \tag{7}$$

Since x^7 is not the zero function, the second factor of (7) must be the zero function. Thus $t(R) = -B/C$. Thus R is the unique point of $\mathcal{X}$ where y has a zero of greater order than x. Note that $(y/x^5)(R) = -B/C$.

We now determine where x and y have poles. This can be done by the use of projective coordinates, however, we will not need them in this paper. Using F and Proposition 5 it is easy to see that the places where x has a pole are precisely the places where y has a pole. Since $deg(x)_\infty = 9$, and $deg(y)_\infty = 7$, we note that there must be a point of $\mathcal{X}$ where x has a pole of greater order than y. By examining F we conclude that Bx^7 has a bigger pole than either Ax^2y^3 or Cx^2y. Thus y^9 and Bx^7 have the same size pole. We denote this point of $\mathcal{X}$ by ∞ and note that $ord_\infty(x) = 9$ and $ord_\infty(y) = 7$. Thus ∞ is the only point of $\mathcal{X}$ where x or y has a pole.

We now find a canonical divisor, in other words, the divisor of a meromorphic differential. All canonical divisors are equivalent, however, some allow for easier computations than others. In practice, if F does not have many singularities, a good divisor to analyze is the following. Since F is the zero function, dF is the zero differential. This yields that $F_x dx + F_y dy = 0$. Define $\hat{\omega} := dx/F_y$ and note that $\hat{\omega} = -dy/F_x$ also. The following proposition is what makes $\hat{\omega}$ useful.

Proposition 28 *Let $G(X,Y)$ be a defining equation for an arbitrary Riemann surface Z. Let V be the locus of points of $\mathbb{C}^2$ which satisfy $G(X,Y) = 0$. Let $(\alpha, \beta) \in V$ be a nonsingular point and let $p \in Z$ correspond to (α, β). Then the differential $\hat{\omega} = dx/G_y = -dy/G_x$ has order 0 at p.*

Proof: If $G_Y(\alpha, \beta) \neq 0$ then $(X - \alpha)$ is not the tangent line, thus Theorem 26 yields that $x - \alpha$ is a local parameter at p. Thus $dx = d(x - \alpha)$ has order 0 at p. Thus dx/G_y has order 0 at p. The argument is analogous for dy/G_x if $G_X(\alpha, \beta) \neq 0$. Thus $\hat{\omega}$ has order 0 at each point of $\mathcal{X}$ which corresponds to a nonsingular point of G.

Since there are few singular points of F, Proposition 28 makes the divisor of $\hat{\omega}$ easy to calculate. As a contrast, let us examine the difficulties one would encounter in trying to calculate the divisor dx. At all points where x does not have a pole, the order of dx is zero except when $F_y = 0$. To determine these points, we would need to determine the intersection of $F = 0$ and $F_y = 0$. This, in general, could be very difficult.

We compute the order of $\hat{\omega}$ at P. From (2) we have that $F_x = x(2Ay^3 + 7Bx^5 + 2Cy)$. Recall that $ord_P(x) = 4$, and $ord_P(y) = 1$, thus $ord_P(dy) = 0$. Note that $ord_P(2Ay^3 + 7Bx^5 + 2Cy) = min\{3, 20, 1\} = 1$, thus $ord_P(F_x) = 5$ and so $ord_P(\hat{\omega}) = ord_P(dy) - ord_P(F_x) = -5$. The exact same calculation yields that $ord_Q(\hat{\omega}) = -5$. At ∞ we have that $ord_\infty(x) = -9$ and $ord_\infty(y) = -7$, therefore $ord_\infty(dy) = -8$ and $ord_\infty(2Ay^3 + 7Bx^5 + 2Cy) = min\{-21, -45, -7\} = -45$. Thus $ord_\infty(F_x) = -54$ and $ord_\infty(\hat{\omega}) = 46$.

The calculation is slightly more complicated at R. Recall that $ord_R(x) = 1$ and $ord_R(y) = 5$. Thus the orders of the terms in $2Ay^3 + 7Bx^5 + 2Cy$ are $15, 5$, and 5 respectively. It is possible that the term $7Bx^5 + 2Cy$ has order greater than 5. However, when we previously examined the point R, we used the change of variables $t = y/x^5$. Thus $ord_R(7Bx^5 + 2Cy) = ord_r(x^5(7B + 2Ct)) = ord_R(x^5) + ord_R(7B + 2Ct)$. Recall that $t(R) = -B/C$, so $ord_R(7B + 2Ct) = 0$. Thus $ord_R(7Bx^5 + 2Cy) = 5 + 0 = 5$, so $ord_R(F_x) = ord_R(x(2Ay^3 + 7Bx^5 + 2Cy)) = 6$ and $ord_R(\hat{\omega}) = ord_R(dy) - ord_R(F_x) = 4 - 6 = -2$.

A similar calculation could have been done if we chose to view $\hat{\omega}$ as dx/F_y. The details are analogous to the above.

We choose to consider the differential $\omega := x\hat{\omega}$ since it has smaller poles in the affine plane. Note that ω has the divisor $(\omega) = -P - Q - R + 37\infty$. We define this divisor to be W.

Recall that the degree of a canonical divisor is $2g - 2$, where g is the genus of the Riemann surface. Since $deg(\omega) = 34$, we obtain that $g = 18$.

We now determine $L(W)$. Note that $L(W)$ is the set of functions which have a zero at P, Q, and R and a pole of order less than 38 at ∞. Recall that the divisors for x and y are $(x) = 4P+4Q+R-9\infty$ and $(y) = P+Q+5R-7\infty$. Thus the following 16 functions are elements of $L(W)$:

$$\{y, y^2, y^3, y^4, y^5, x, xy, xy^2, xy^3, xy^4, x^2, x^2y, x^2y^2, x^3, x^3y, x^4\}.$$

It is easy to see that they are independent over $\mathbb{C}$ since they have distinct orders at ∞. Since there are $g = 18$ independent elements of $L(W)$, we search for two others. It is easy to see that $(y^5/x) = P + Q + 24R - 26\infty$ and $(y^6/x) = 2P + 2Q + 29R - 33\infty$. Thus $L(W)$ is spanned by the above 16 elements and y^5/x and y^6/x. We note, in particular, that $L(W)$ contains every polynomial in x and y without constant term of total degree less than or equal to four. Note also that if $f \in L(W)$, then $f\omega$ is a holomorphic differential. Thus we have found a basis for the holomorphic differentials of $\mathcal{X}$.

Since $\mathcal{X}$ has genus 18 and $(x)_\infty = 9\infty$ and $(y)_\infty = 7\infty$, we see that ∞ is a Weierstrass point of $\mathcal{X}$. We now show that every other point of $\mathcal{X}$ has either 7 or 9 as a gap.

Let $(\alpha, \beta) \neq (0,0)$ be any point of V and let T be the corresponding point of $\mathcal{X}$. To show 7 is a gap at T, Proposition 27 yields it is sufficient to find an $f \in L(W)$ such that $ord_T(f) = 6$. Similarly, to show 9 is a gap at T, we find an $f \in L(W)$ such that $ord_T(f) = 8$.

We divide our arguments into several cases:

1. Suppose $ord_T(x - \alpha) > 1$ and $ord_T(y - \beta) = 1$. If $ord_T(x - \alpha) = 2, 3, 4, 5$, or 6, then we may choose f to be $(x - \alpha)^3x, (x - \alpha)^2x, (x - \alpha)(y - \beta)^2x$, $(x - \alpha)(y - \beta)x$, or $(x - \alpha)x$. Each of these functions is in $L(W)$ and has a zero of order 6 at T. If $ord_T(x - \alpha) = 7$, then $y(y - \beta)(x - \alpha)$ has order 8. If $ord_T(x - \alpha) = 8$, then $y(x - \alpha)$ has order 8. Thus 7 or 9 are gaps at T. Finally, we note that $(x - \alpha)$ cannot have order 9 at T. If this were so, then $F(\alpha, Y)$ would have exactly one root, thus $F(\alpha, Y) = (Y - \beta)^9$. Clearly this is impossible, since F has no term of degree 8 in y.

2. Suppose $ord_T(x - \alpha) = 1$ and $ord_T(y - \beta) > 1$. If $ord_T(y - \beta)$ is 2,3,4,5, or 6, then interchanging the role of $x - \alpha$ and $y - \beta$ in the previous case will yield functions in $L(W)$ with a zero of order 6. If $ord_T(y - \beta) = 7$, then $x(x - \alpha)(y - \beta)$ has order 8. Thus 7 or 9 are gaps at T.

3. Finally, suppose $ord_T(x - \alpha) = 1$ and $ord_T(y - \beta) = 1$. In this case, there exists a nonzero complex number γ such that $ord_T(x - \alpha + \gamma(y - \beta)) > 1$. Define $h = x - \alpha + \gamma(y - \beta)$. If $ord_T(h) = 2, 3, 4, 5$, or 6, then we may choose f to be $h^3x, h^2x, h(y - \beta)^2x, h(y - \beta)x$, or hx respectively. In each case, f has a zero of order 6 at T. If $ord_T(h) = 7$, then $ord_T(x(x - \alpha)h) = 8$, while if $ord_T(h) = 8$, then $ord_T(xh) = 8$. Thus 7 or 9 are gaps at T. We see that it is impossible that $ord_T(h) = 9$ from the following. If $ord_T(h) = 9$, then in Equation 2, we rewrite F in terms of y and h by making the change of variables $x = \alpha + h - \gamma(y - \beta)$, and call this new polynomial $\hat{F}(h, y)$. Thus there is only one point of $\mathcal{X}$ where h has a

zero. Thus $\hat{F}(0, y) = (y - \beta)^9$. This is a contradiction, since $\hat{F}$ has no term of degree 8 in y.

To determine that 7 is a gap at P, Q and R, we use the modification stated after Proposition 27. At each point we must find a function in $L(W)$ with a zero of order exactly 7. At P and Q, the function xy^3 has a zero of order exactly 7, while at R the function x^2y has a zero of order exactly 7. Thus 7 is a gap at P, Q or R.

We have shown that ∞ is a distinguished Weierstrass point; there is no other point of $\mathcal{X}$ with the same gap sequence as at ∞. Thus any automorphism of $\mathcal{X}$ must map ∞ to itself. This greatly restricts the automorphism group of $\mathcal{X}$ and allows us to compute it explicitly.

Let σ be an automorphism of $\mathcal{X}$. We will denote the induced automorphism of $\mathbb{C}(x, y)$ by σ also. Any function which has a pole only at ∞ must be mapped to a function with a pole only at ∞. However, given what we know about $\mathcal{X}$, it is easy to determine all functions which have a pole precisely at ∞. Note that $l((2g - 1)\infty) = g$, therefore, $l(35\infty) = 18$. Note that $L(35\infty)$ includes all of the elements of $L(W)$ whose pole at ∞ is of order 35 or less, as well as 1 and y^4/x.

Since $(y)_\infty = 7\infty$, $(\sigma(y))_\infty = 7\infty$. Similarly, since $(x)_\infty = 9\infty$, $(\sigma(x))_\infty = 9\infty$. Therefore $\sigma(y) = a + by$, and $\sigma(x) = c + dy + ex$ for some complex numbers a, b, c, d and e. Thus

$$0 = \sigma(F) =$$

$$(a + by)^9 + A(c + dy + ex)^2(a + by)^3 + B(c + dy + ex)^7 + C(c + dy + ex)^2(a + by).$$

Since F has no term of degree 8 in y, the uniqueness of the minimal polynomial of y over $C(x)$ yields that $a = 0$. Since F has no term of degree 6 in x or 7 in y, we obtain that $c = 0$ and $d = 0$. Thus $\sigma(x) = ex$ and $\sigma(y) = by$. But then $0 = y^9 + (Ae^2b^3/b^9)x^2y^3 + B(e^7/b^9)x^7 + C(e^2b/b^9)x^2y$. Thus $e^2/b^6 = 1$, $e^7/b^9 = 1$, and $e^2/b^8 = 1$. This implies that σ is the identity or $\sigma(x) = -x$ and $\sigma(y) = -y$. Thus the full automorphism group of $\mathcal{X}$ has order two.

To see how the above techniques are used to construct an explicit family of Riemann surfaces which possess trivial automorphism groups, see [8].

References

[1] Lars V. Ahlfors, *Complex Analysis*, McGraw Hill: New York, New York, 1979.

[2] E. Bujalance, J. J. Etayo, J.M. Gamboa, G. Gromadzki, *Automorphisms of Compact Bordered Klein Surfaces. A Combinatorial Approach*. Lecture Notes in Math. vol. 1439, Springer Verlag: New York, New York, 1990.

[3] Claude Chevalley, *Introduction to the Theory of Algebraic Functions of One Variable,* Amer. Math. Soc. Survey, No. 6, American Mathematical Society: Providence, Rhodé Island, 1951.

[4] H.M. Farkas and I. Kra, *Riemann Surfaces,* Graduate Texts in Mathematics No. 71, Springer Verlag: New York, New York, 1992.

[5] William Fulton, *Algebraic Curves,* Benjamin: New York, New York, 1969.

[6] R. C. Gunning, *Lectures on Riemann Surfaces*, Princeton University Press: Princeton, New Jersey, 1966.

[7] George Springer *Introduction to Riemann Surfaces*, Addison Wesley: Reading, Massachusetts, 1957

[8] P. Turbek, *An explicit family of curves with trivial automorphism groups,* Proceed. Amer. Math. Soc., 122, No. 3, (1994).

Department of Mathematics, Comp. Sci., and Statistics,
Purdue University Calumet,
Hammond, Indiana, 46323, USA
turbek@nwi.calumet.purdue.edu

SYMMETRIES OF RIEMANN SURFACES FROM A COMBINATORIAL POINT OF VIEW

Grzegorz Gromadzki

1. On combinatorial approach in studies of automorphisms of Riemann surfaces

Throughout the lecture a Riemann surface is meant to be a compact Riemann surface of genus $g \geq 2$ and a *symmetry* of such surface, an antiholomorphic involution. The reader who has attended the lecture [11] must be convinced of the importance of studies of symmetries of Riemann surfaces and the role that the groups of their automorphisms play there. Up to certain extent that lecture illustrates how to use results concerning this subject and in this one we shall mainly show how to get them. A symmetric Riemann surface X corresponds to a complex curve $\mathcal{C}_X$ which can be defined over the reals and symmetries nonconjugated in the group $\mathrm{Aut}^{\pm}(X)$ of all automorphisms of X correspond to nonequivalent real forms of $\mathcal{C}_X$. The aim of this lecture is a brief introduction to the combinatorial aspects of this theory together with samples of results and proofs. The most natural questions that arise here are the following:

- does a Riemann surface X admit a symmetry?
- how many nonconjugated symmetries may a given Riemann surface X admit?
- what can one say about the topology of these symmetries, e.g. about the number of their ovals or about their separability?

Riemann surfaces form a category with the holomorphic maps as morphisms and this category is closed under the quotients with respect to the action of groups of holomorphic automorphisms. In the context of symmetries it is convenient to enlarge the class of morphisms taking into account antiholomorphic maps which roughly speaking are those ones that admit in their local forms analytic maps followed by the complex conjugation. However now it is no more true that such category is closed under quotients with respect to groups containing also antiholomorphic automorphisms; the so called *Klein surfaces* appear as new objects [2].

The starting point in studies of such surfaces from a combinatorial point of view is the Riemann uniformization theorem that we are going to explain shortly. Given a Riemann surface X of genus g forget, for the moment, about its analytic structure and consider the universal covering $p : \tilde{X} \to X$. Then $\tilde{X}$ is a simply connected topological surface which can be made a Riemann surface, simply by lifting the analytic structure on X; p becomes in such situation a holomorphic map. Now the mentioned theorem of Riemann says that, up to holomorphic isomorphism, $\tilde{X}$ is either the sphere $C \cup \{\infty\}$ or the

Euclidean plane C or else the upper half plane $\mathcal{H} = C^+$ according to whether $g = 0, 1$ or $g \geq 2$ respectively. The sphere is the unique Riemann surface of genus 0. Here we shall deal with surfaces of genus $g \geq 2$; Riemann surfaces of genus 1 correspond to so called elliptic curves and they govern themselves by their own rules and methods, rather different to those concerning surfaces of genus $g \geq 2$.

So, for a Riemann surface X, we always have an unramified holomorphic map $p_X : \mathcal{H} \to X$. Consider the group Γ of its deck transformations i.e. the group of all homeomorphisms of $\mathcal{H}$ which commute with p_X. Its elements turn out to be automorphisms of $\mathcal{H}$. Let $\mathcal{G}$ be the group of all such automorphisms. It is often convenient to have in mind that elements of $\mathcal{G}$ can be also considered as isometries of $\mathcal{H}$ viewed as the hyperbolic plane. Furthermore, it is frequently useful to use the open unit disc $\mathcal{D}$ as a model of such plane. We shall use alternatively both models and in this section we shall consider the action of $\mathcal{G}$ as a right one. As an abstract group, Γ is isomorphic to the fundamental group of X. From the general theory of covering spaces, p_X induces a homeomorphism $\tilde{p}_X$ between the orbit space $\mathcal{H}/\Gamma$ and X. This homeomorphism introduces on $\mathcal{H}/\Gamma$ a structure of Riemann surface and it turns out that this is the unique structure under which the canonical projection $\mathcal{H} \to \mathcal{H}/\Gamma$ is a holomomorphic map. One usually says that Γ *uniformizes* X. Furthermore if G is a group of automorphisms of X then all liftings of all elements of G form a group Δ of automorphisms of $\mathcal{H}$ for which $G \cong \Delta/\Gamma$. Conversely each such quotient can be viewed as a group of automorphisms of $\mathcal{H}/\Gamma$.

The groups Γ and Δ are Fuchsian groups, i.e. discontinuous and cocompact subgroups of orientation preserving isometries of $\mathcal{H}$ and the point is that the algebraic structure of such groups is well known; cocompact means that the corresponding orbit spaces are compact. One usually obtains a presentation of a Fuchsian group Δ by means of the defining generators and relators by using of fundamental region F and we shall shortly describe this procedure. Since Δ is discontinuous it follows that there exists $p \in \mathcal{D}$ with a trivial stabilizer in Γ. Now given $\delta \in \Delta$,

$$F_\delta^p = \{z \in \mathcal{D} \ : \ d(z,p) \leq d(z,p\delta)\}$$

is the semiplane containing p and determined by the perpendicular bisector of the hyperbolic segment $[p, p\delta]$. The cocompactness of Δ guarantees that a finite number of these semiplanes cover $\mathcal{D}$ and therefore

$$F = F_p = \bigcap_{\delta \in \Delta} F_\delta^p$$

is a finitely edged convex polygon which have two properties:
  (a) given $z \in \mathcal{D}$ there exists $\delta \in \Delta$ for which $z\delta \in F$,
  (b) the above δ is unique if $z\delta \in \text{Int}(F)$.

Each convex region F bounded by a finitely edged polygon and satisfying the above conditions is said to be a fundamental region for Δ; the above constructed one is called the *Dirichlet fundamental region* with center at p. Its images $F\delta$ are said to be the *faces* of F. Clearly the map

$$\delta \mapsto F\delta$$

is a bijection beetwen Δ and the set of faces of F. Call a sequence $F_1, \ldots, F_n$ of faces to be a *chain* if F_{i-1} meets F_i in an edge, for each $i \leq n$. Given an edge of F, say labelled by e, there is a unique transformation δ_e for which $F\delta_e$ meets F along e. Observe that for δ_e so defined, $e'\delta_e = e$ if and only if $\delta_{e'} = \delta_e^{-1}$; such edges e and e' are said to be *congruent*. We claim that the set $\{\delta_e\}$ for e running over the edges of F generate Δ. For, observe first that given δ and δ' for which $F\delta$ meets $F\delta'$ in a common edge, $\delta_e\delta = \delta'$ for $e = F \cap F\delta'\delta^{-1}$ (see Figure 1). Now given $\delta \in \Delta$ there is a chain $F = F_0, F_1, \ldots, F_n = F\delta$. Then $F_i = F\delta_i$ for some $\delta_i \in \Delta$. So $\delta_{i+1} = \delta_{e_{i+1}}\delta_i$ and therefore $\delta = \delta_{e_n} \ldots \delta_{e_1}$ for some edges $e_1, \ldots, e_n$ of F. One usually says that Δ is generated by the edges of its fundamental region. We shall call these generators, the *canonical* (*with respect to F*) ones.

$$F\delta_e = F\delta'\delta^{-1}$$

If $F_n = F\delta$, then $\delta = \delta_{e_1} \ldots \delta_{e_n}$, where $F_i\delta_{e_{i+1}} = F_{i+1}, i = 1, \ldots, n-1$.

Figure 1

Now let $F = F_0, F_1, \ldots, F_{m-1}, F_m = F$ be the chain of all consecutive faces meeting F at a vertex ν written in the clockwise order. Then $F_i = F_{i-1}\delta_i$ for some $\delta_i \in \Delta$ and so the product $\delta_1 \ldots \delta_m$ is a relator which can be written as a word in canonical generators rewriting each δ_i in such terms. We claim that these relators, which again will be called *canonical* (*with respect to F*), form a set of defining relators for Δ. For, observe first that we have a bijective correspondence between chains starting and ending at F and relators with respect to canonical generators:

$$r = \delta_{e_1} \ldots \delta_{e_{n+1}} \longleftrightarrow F = F_0, F_1, \ldots, F_n, F_{n+1} = F,$$

where $F_i = F\delta_{e_i} \ldots \delta_{e_1}$. Call a chain $F = F_0, F_1, \ldots, F_n, F_{n+1} = F$ to be *simple* if $F_0 = F_{n+1}, \ldots, F_s = F_{n+1-s}$ and $F_{s+1}, \ldots, F_{n-s}$ are distinct for some s but have a common point. Observe that, under the above bijection, simple chains correspond to relators conjugated to canonical. Indeed if $\delta = \delta_1 \ldots \delta_s$ then $F_i \delta^{-1}, i = s + 1, \ldots, n - s$ is the chain of all faces meeting $F = F_s \delta^{-1}$ at some vertex and, up to trivial cancellations $\delta_e \delta_e^{-1} = \emptyset$, $\delta^{-1} r \delta$ is a canonical relator. We shall show that each relator is a product of relators conjugated to canonical ones. For, let

$$F = F_0, F_1, \ldots, F_n, F_{n+1} = F$$

be a chain of faces corresponding to a relator $r = \delta_{e_1} \ldots \delta_{e_{n+1}}$. We shall prove the assertion by induction on the number m of vertices of the lattice corresponding to F which are in the interior of the region bounded by this chain. (For the first three chains from Figure 2, $m = 15, 1$ and 14 respectively.)

For $m = 0$ the relator is trivial while for $m = 1$, up to trivial cancellations, it is a power of a canonical relator. So let $m > 1$. Then inside of a region bounded by our chain we have a vertex $\nu \notin F$ being a common vertex of some F_i and F_{i+1} and in addition F_{i+1} is an exterior face. Let $G_1, \ldots, G_m$ be the chain of all faces meeting ν, where $G_1 = F_i$ and $G_2 = F_{i+1}$ (observe that some other, but not necesarrily all G_i's, can appear in the former chain). Then

and
$$F_0, \ldots, F_{i-1}, G_1, \ldots, G_m, F_i, \ldots, F_0$$
$$F_0, \ldots, F_i, G_m, \ldots, G_3, F_{i+2}, \ldots, F_n$$

are two chains which give relators r_1, r_2 for which $r = r_1 r_2$. The first chain is simple and so the corresponding relator is conjugate to a canonical one. Repeating this procedure for the second chain as many times as F_{i+1} appears in the former chain we can assume that r_2 corresponds to a chain having less than m vertices in the interior of the region that it bounds. Therefore the assertion follows. This process is illustrated in Figure 2 in the next page for three typical configurations.

Now the fundamental region F can be modified, by the so called *cutting and gluing* technique so that the corresponding system of canonical generators gives certain canonical presentation. To explain it generally, consider a pair (e, e') of congruent sides of F end let ℓ be a hyperbolic segment joining two vertices of F and dividing F into two subregions F_1 and F_2 that contain edges e' and e respectively. Then $F' = F_2 \cup F_1 \delta_e$ is a fundamental region for Δ whose edges are identified by suitable elements of Δ.

Now the fundamental region F can be modified, by the so called *cutting and gluing* technique so that the corresponding system of canonical generators gives certain canonical presentation. To explain it generally, consider a pair (e, e') of congruent sides of F end let ℓ be a hyperbolic segment joining two vertices of F and dividing F into two subregions F_1 and F_2 that contain edges e' and e respectively. Then $F' = F_2 \cup F_1 \delta_e$ is a fundamental region for Δ whose edges are identified by suitable elements of Δ.

Notice first that the canonical generators can be divided into two classes: those which have fixed points called *elliptic* elements or *hyperbolic rotations* and those without fixed points called *hyperbolic* elements or *hyperbolic translations*. Observe that *parabolic* elements (i.e, those which have single fixed points at infinity) does not appear here since Δ is cocompact.

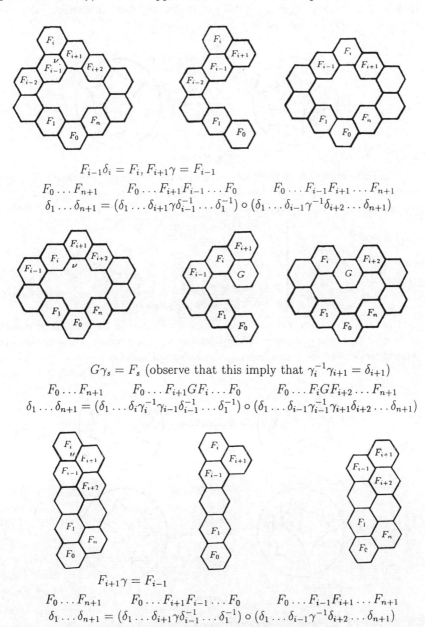

$$F_{i-1}\delta_i = F_i, F_{i+1}\gamma = F_{i-1}$$

$$
\begin{array}{ccc}
F_0 \dots F_{n+1} & F_0 \dots F_{i+1}F_{i-1}\dots F_0 & F_0 \dots F_{i-1}F_{i+1}\dots F_{n+1}
\end{array}
$$

$$\delta_1 \dots \delta_{n+1} = (\delta_1 \dots \delta_{i+1}\gamma\delta_{i-1}^{-1} \dots \delta_1^{-1}) \circ (\delta_1 \dots \delta_{i-1}\gamma^{-1}\delta_{i+2} \dots \delta_{n+1})$$

$$G\gamma_s = F_s \text{ (observe that this imply that } \gamma_i^{-1}\gamma_{i+1} = \delta_{i+1})$$

$$
\begin{array}{ccc}
F_0 \dots F_{n+1} & F_0 \dots F_{i+1}GF_i \dots F_0 & F_0 \dots F_iGF_{i+2}\dots F_{n+1}
\end{array}
$$

$$\delta_1 \dots \delta_{n+1} = (\delta_1 \dots \delta_i\gamma_i^{-1}\gamma_{i-1}\delta_{i-1}^{-1} \dots \delta_1^{-1}) \circ (\delta_1 \dots \delta_{i-1}\gamma_{i-1}^{-1}\gamma_{i+1}\delta_{i+2} \dots \delta_{n+1})$$

$$F_{i+1}\gamma = F_{i-1}$$

$$
\begin{array}{ccc}
F_0 \dots F_{n+1} & F_0 \dots F_{i+1}F_{i-1}\dots F_0 & F_0 \dots F_{i-1}F_{i+1}\dots F_{n+1}
\end{array}
$$

$$\delta_1 \dots \delta_{n+1} = (\delta_1 \dots \delta_{i+1}\gamma\delta_{i-1}^{-1} \dots \delta_1^{-1}) \circ (\delta_1 \dots \delta_{i-1}\gamma^{-1}\delta_{i+2} \dots \delta_{n+1})$$

Figure 2

Elliptic elements identify two consecutive edges which we can label by x_i and x_i'. Operating as illustrates Figure 3 we can assume that all elliptic generators correspond to edges which form a connected segment of the boundary. Here and in the next figures capital letters denote blocks of edges; X denotes the block of edges corresponding to elliptic elements.

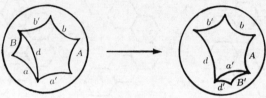

Figure 3

It is rather instructive the analysis of how the canonical generators δ_f' given by the new fundamental region F' are related to the canonical generators δ_e corresponding to the former fundamental region F. The reader can check that

$$\delta_f' = \begin{cases} \delta_f & \text{if } f = b \text{ or } b', \\ \delta_a & \text{if } f = d, \\ \delta_{a'} & \text{if } f = d', \\ \delta_f & \text{if } f \in A \text{ and } f' \in A, \\ \delta_a \delta_f & \text{if } f \in A \text{ and } f' \in B, \\ \delta_e \delta_a^{-1} & \text{if } f \in B', \text{ where } e = f\delta_d. \end{cases}$$

Now using a composition of operations illustrated in Figure 4 (the reader can easily find how generators corresponding to a new fundamental region are related to generators corresponding to the former region)

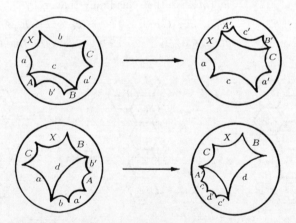

Figure 4

we arrive to a fundamental region whose edges can be labelled as

$$x_1 x_1' \ldots x_r x_r' a_1 b_1 a_1' b_1' \ldots a_g b_g a_g' b_g'$$

For the sake of simplicity, denote the canonical generators by the same symbol as the edges to which they correspond. Observe that with such convention $e' = e^{-1}$. Now each vertex between x_i and x'_i, gives a relator of the form $x_i^{m_i}$ and each other vertex gives a relator obtained by a cyclic permutation of a relator of the form

$$x_1 \dots x_r a_1 b_1 a_1^{-1} b_1^{-1} \dots a_g b_g a_g^{-1} b_g^{-1}.$$

So if we take one generator for each pair of congruent edges, we obtain the following canonical presentation of Δ:

(1) $\quad \langle x_1, \dots, x_r, a_1, b_1, \dots, a_{g'} b_{g'} \mid x_1^{m_1}, \dots, x_r^{m_r}, \prod_{i=1}^{r} x_i \prod_{i=1}^{g'} a_i b_i a_i^{-1} b_i^{-1} \rangle$

for some sequence of integers

(2) $\qquad\qquad\qquad\qquad (g'; m_1, \dots, m_r)$

called the *signature* of Δ.

The Fuchsian group Γ uniformizing the Riemann surface X acts on $\mathcal{H}$ without fixed points. So it has signature $(g; -)$ and if $G = \Delta/\Gamma$, then the signature $(g'; m_1, \dots, m_r)$ of Δ provides the topological features of the action of G on X; X/G has genus g' and $m_1, \dots, m_r$ are the ramification data of the the canonical projection $X \to X/G$.

The hyperbolic area $\mu(\Delta)$ of an arbitrary fundamental region of a Fuchsian group Δ with signature (2) is, by the Gauss-Bonnet formula, equal to

(3) $\qquad\qquad\qquad 2\pi \left(2g' - 2 + \sum_{i=1}^{r} \left(1 - \frac{1}{m_i} \right) \right).$

Furthermore if Δ' is a subgroup of Δ of finite index then it is also a Fuchsian group and we have the following Hurwitz Riemann formula:

(4) $\qquad\qquad\qquad\qquad [\Delta : \Delta'] = \mu(\Delta')/\mu(\Delta)$

Now the point is that an abstract group with the presentation (1) can be realized as a Fuchsian group if and only if (3) is positive. This is what actually makes possible combinatorial studies of automorphism groups of Riemann surfaces. Indeed, let (δ, γ) be a pair formed by an abstract group δ and by its normal subgroup γ with the presentations corresponding to signatures $(g'; m_1, \dots, m_r)$ and $(g; -)$ respectively, for which the values (3) are positive. Then there exists a Fuchsian group Δ isomorphic to δ. Such Δ contains as a normal subgroup a Fuchsian group Γ isomorphic to γ and so $\delta/\gamma \cong \Delta/\Gamma$ can be represented as a group of automorphisms of $\mathcal{H}/\Gamma$ which is a Riemann surface of genus g. We see therefore that, stated in such a way, the problem of studies of Riemann surfaces and their groups of automorphisms is actually a purely algebraic task where, as we shall see, the conformal aspects play a game just up to some extent only.

2. Symmetries, symmetric Riemann surfaces and their groups of automorphisms

The constructions from the previous section concern holomorphic automorphisms and here we shall explain how studies of symmetries of Riemann surfaces fit in this general algebraically combinatorial scheme. Assume that X is a Riemann surface of genus $g \geq 2$, uniformized by a Fuchsian surface group Γ. Then the group $\mathrm{Aut}^+(X)$ of all holomorphic automorphisms of X can be represented as the factor group Δ/Γ for some other Fuchsian group Δ and there are no Fuchsian group containing Δ as a proper subgroup and at the same time Γ as a normal subgroup. Let $p : \mathcal{H} \to X$ be the canonical covering. Then a symmetry σ of X can be lifted to a homeomorphism λ of $\mathcal{H}$ which composed with the reflection $z \mapsto -\bar{z}$ of $\mathcal{H}$ is a sense-preserving isometry and therefore λ is a reflection or glide reflection of $\mathcal{H}$. Observe now that λ normalizes Δ and Γ and so together with Δ generate a group Λ containing Δ as a subgroup of index 2 and Γ as a normal subgroup. So $G = \Lambda/\Gamma = \mathrm{Aut}^\pm(X)$ is the group of all automorphisms of X.

The above group Λ is an example of an NEC group. Generally, *NEC groups* are defined as discontinuous subgroups of isometries of $\mathcal{H}$ with compact orbit space $\mathcal{H}/\Lambda$. Again the algebraic structure of these groups is well known. Their presentations, by means of defining generators and relators are determined by signatures $\sigma(\Lambda)$ being sequences of symbols and numbers of the form

$$(5) \qquad (g'; \pm; [m_1, \ldots, m_r]; \{C_1, \ldots, C_n\}),$$

where the numbers $m_i \geq 2$ are called the *proper periods*, $C_i = (n_{i1}, \ldots, n_{is_i})$ are called the *period cycles*, the numbers $n_{ij} \geq 2$ are the *link periods* and $g' \geq 0$ is said to be the *orbit genus* of Λ [20], [30]. A group Λ with signature (5) has the following presentation:

(6)
$$
\left\{
\begin{array}{l}
\text{generators:} \\[4pt]
\quad \text{(i)} \quad x_i,\ i = 1, \ldots, r, \\[2pt]
\quad \text{(ii)} \quad c_{ij},\ i = 1, \ldots, k,\ j = 0, \ldots, s_i, \\[2pt]
\quad \text{(iii)} \quad e_i,\ i = 1, \ldots, k, \\[2pt]
\quad \text{(iv)} \quad a_i,\ b_i,\ i = 1, \ldots, g \text{ if the sign is } +, \\[2pt]
\qquad\quad d_i,\ i = 1, \ldots, g, \text{ if the sign is } -, \\[4pt]
\text{relations:} \\[4pt]
\quad \text{(i)} \quad x_i^{m_i} = 1,\ i = 1, \ldots, r, \\[2pt]
\quad \text{(ii)} \quad c_{is_i} = e_i^{-1} c_{i0} e_i,\ i = 1, \ldots, k, \\[2pt]
\quad \text{(iii)} \quad c_{ij-1}^2 = c_{ij}^2 = (c_{ij-1} c_{ij})^{n_{ij}} = 1,\ i = 1, \ldots, k,\ j = 1, \ldots, s_i, \\[2pt]
\quad \text{(iv)} \quad x_1 \ldots x_r e_1 \ldots e_k [a_1, b_1] \ldots [a_g, b_g] = 1, \text{ if the sign is } +, \\[2pt]
\qquad\quad x_1 \ldots x_r e_1 \ldots e_k d_1^2 \ldots d_g^2 = 1 \text{ if the sign is } -.
\end{array}
\right.
$$

Crucial role in studies of symmetries of Riemann surfaces play generators c_{ij} which represent reflections; reflections c_{ij-1} and c_{ij} are said to be *consecutive*. We shall see in the next section that each reflection of Λ is conjugate in Λ to some c_{ij}. NEC groups with signatures $(g'; \pm; [-]; \{(-), .\overset{k}{.}., (-)\})$ are called *surface NEC groups* and uniformize Klein surfaces of topological genus g' with k boundary components and orientable or not according to whether the sign is $+$ or $-$. A Fuchsian group can be regarded as an NEC group with signature $(g'; +; [m_1, \ldots, m_r]; \{-\})$ and then a Fuchsian surface group is an NEC group with signature $(g'; +; [-]; \{-\})$.

Presentation (6) for NEC groups can be found in a similar manner as for Fuchsian groups; one must however keep in mind that reflections and glide reflections are involved. This makes the matter a little bit more complicated since now we have two more classes of edges: those which are not identified with any other edge and those which are identified with some other edge by orientation reversing transformations; label the first ones by c's and the pairs of congruent edges of the second type by (d, d^*). Then after suitable modifications of the initial fundamental region we arrive to a fundamental region labelled as

$$x_1 x_1' \ldots x_r x_r' e_1 c_{10} \ldots c_{1s_1} e_1' \ldots e_k c_{k0} \ldots c_{ks_k} e_k' a_1 b_1 a_1' b_1' \ldots a_g b_g a_g' b_g'$$

or

$$x_1 x_1' \ldots x_r x_r' e_1 c_{10} \ldots c_{1s_1} e_1' \ldots e_k c_{k0} \ldots c_{ks_k} e_k' d_1 d_1^* \ldots d_g d_g^*$$

according to $\mathcal{H}/\Lambda$ is orientable or not. From each of them it is easy to obtain the above presentation.

Similar, as for Fuchsian groups, facts concerning metric aspects of the fundamental regions hold. The hyperbolic area of an arbitrary fundamental region of an NEC group Λ with signature (5) equals:

$$(7) \qquad \mu(\Lambda) = 2\pi \left(\varepsilon g' - 2 + k + \sum_{i=1}^{r} \left(1 - \frac{1}{m_i} \right) + \frac{1}{2} \sum_{i=1}^{k} \sum_{j=1}^{s_i} \left(1 - \frac{1}{n_{ij}} \right) \right),$$

where $\varepsilon = 2$ if there is a $"+"$ sign and $\varepsilon = 1$ otherwise, and the Hurwitz Riemann formula $[\Lambda : \Lambda'] = \mu(\Lambda')/\mu(\Lambda)$ holds.

3. Subgroups of NEC groups

At the beginning of the eighties, a method (based mainly on "surgery" of a fundamental region) of determining the signature of a normal subgroup Λ of an NEC group Λ' as a function of the quotient group Λ'/Λ and the signature $\sigma(\Lambda')$ of Λ' was developed by E. Bujalance [5], [6]. For our purposes however, we shall need a description of $\sigma(\Lambda)$ when Λ is not necessarily normal in Λ'. Hoare [17] has developed an algorithm to find in such situation proper periods and period cycles of Λ. Unfortunately its computational complexity restricts essentially the area of effective calculations for large indices. An advantage of our situation will be that we shall know a priori that $\sigma(\Lambda)$ has a rather special form and in section 6 we shall present an alternative approach. Here and later we shall need the following technical lemma.

Lemma 1 *Any reflection of an NEC group is conjugate to one of its canonical reflections and any elliptic element is conjugate to a power of some of its canonical elliptic generators or to a power of the product of two consecutive canonical reflections.*

Proof. Indeed if ℓ is the axis of a reflection $c' \in \Lambda$, then there exists $\lambda \in \Lambda$ for which $\lambda(\ell)$ meets an edge of a given fundamental region F. So since $\lambda(\ell)$ is the set of fixed points of $\lambda c' \lambda^{-1}$, the last is a canonical reflection. Similarly if p is the fixed point of an elliptic element x, then $\lambda(p) \in F$ for some $\lambda \in \Lambda$ and on the other hand it is the fixed point of $x^\lambda = \lambda x \lambda^{-1}$. Therefore $\lambda(p)$ is a vertex of F and x^λ is a power of some canonical elliptic element of Λ or a power of the product of two consecutive canonical reflections.

The next theorem, proved by Singerman [28], is very important in studies of symmetric Riemann surfaces since, up to certain extent, it describes how the symmetry type of a Riemann surface X depends on the topological characteristics of the action of the group $\mathrm{Aut}^+(X)$ of all holomorphic automophisms of X.

Theorem 2 *The canonical Fuchsian subgroup Λ^+ of an NEC group Λ with signature (5) has signature $(\varepsilon g + k - 1; m_1, m_1, \ldots, m_r, m_r, n_{11}, \ldots, n_{ks_k})$, where $\varepsilon = 2$ if the sign of Λ is $+$ and $\varepsilon = 1$ otherwise.*

Proof. We have to look for conjugacy classes in Λ^+ of elliptic elements of Λ. Let λ be an arbitrary element of $\Lambda \setminus \Lambda^+$. Then, each canonical elliptic element x_i of Λ and its conjugate x_i^λ belong to Λ^+ and they are not conjugated there. Indeed let p be the fixed point of x_i and assume that $x_i^\gamma = x_i^\lambda$, for some $\gamma \in \Lambda^+$. Then $\gamma^{-1}\lambda$ has p as a fixed point and therefore $\gamma^{-1}\lambda$ is a reflection with axis passing by p, a contradiction since in such case $\gamma^{-1}\lambda = x_i^m$ for some m. Furthermore for arbitrary $\delta \in \Lambda$, x_i^δ is conjugate in Λ^+ to x_i or to x_i^λ according to δ belongs to Λ^+ or not. Finally $c_{ij-1}c_{ij} \in \Lambda^+$ and for any $\lambda \in \Lambda \setminus \Lambda^+$, $(c_{ij-1}c_{ij})^\lambda = (c_{ij}c_{ij-1})^{\lambda c_{ij-1}}$. The orbit genus of Λ^+ can be found using the Hurwitz–Riemann formula (4). So the result follows.

The next result can be derived from Singerman's thesis [26] (see also [28]).

Theorem 3 *Let $c_0, c_1, \ldots, c_s, e$ be the system of canonical generators corresponding to a period cycle $(n_1, \ldots n_s)$ of an NEC group Λ with respect to a fundamental region F. If all n_i are even then the centralizer $\mathrm{C}(\Lambda, c_i)$ of c_i in Λ equals*

$$\langle c_i \rangle \oplus \left(\langle (c_{i-1}c_i)^{n_i/2} \rangle * \langle (c_i c_{i+1})^{n_{i+1}/2} \rangle \right) = \mathbb{Z}_2 \times (\mathbb{Z}_2 * \mathbb{Z}_2) \quad \textit{if } i \neq 0,$$

$$\langle c_0 \rangle \oplus \left(\langle (ec_s e^{-1} c_0)^{n_s/2} \rangle * \langle (c_0 c_1)^{n_1/2} \rangle \right) = \mathbb{Z}_2 \times (\mathbb{Z}_2 * \mathbb{Z}_2) \quad \textit{if } i = 0,$$

$$\langle c_0 \rangle \times \langle e \rangle = \mathbb{Z}_2 \times \mathbb{Z} \quad \textit{if } s = 0.$$

Proof. Observe that the inclusions $\supseteq$ are obvious. We shall prove the converse. For $i \neq 0$, let $\gamma_{i-1}, \gamma_i, \gamma_{i+1}$ be the edges of F corresponding to c_{i-1}, c_i, c_{i+1}. Let ℓ be the hyperbolic line containing γ_i. Then for $\lambda \in \Lambda$, $\lambda c_i \lambda^{-1}$ is a reflection with axis $\lambda(\ell)$. So λ centralizes c_i if and only if $\lambda(\ell) = \ell$, which is equivalent to the fact that $\lambda(F)$ is adjacent to ℓ. Let $|\gamma_i|$ be the hyperbolic length of γ_i. Then $(c_{i-1}c_i)^{n_i/2}(c_ic_{i+1})^{n_{i+1}/2}$ is a translation along ℓ with length $2|\gamma_i|$, since it is the composition of two half turns with respect to the ends of γ_i. Now composing certain power of $(c_{i-1}c_i)^{n_i/2}(c_ic_{i+1})^{n_{i+1}/2}$ with $c_i(c_ic_{i+1})^{n_{i+1}/2}$ we can produce an element $\lambda \in \Lambda$ such that $\lambda(F)$ can be an arbitrary face adjacent to ℓ and lying on the same side of ℓ as F while composing the last with c_i we can produce an element $\lambda \in \Lambda$ such that $\lambda(F)$ can be an arbitrary face adjacent to ℓ and lying on the other side of ℓ as F. The case $i = 0$ is similar; actually one must repeat the above arguments for the triple of reflections $e^{-1}c_s e, c_0, c_1$.

For $s = 0$, e fixes ℓ. So F and $e(F)$ have a common edge and both of them are adjacent to ℓ. So $e^k(F)$ can be an arbitrary face adjacent to ℓ and lying on the same side of ℓ as F while considering $c_0 e^k(F)$ we can obtain an arbitrary face lying on the other side of ℓ. Therefore when k runs over all integers and $\varepsilon = 0$ or 1, $c_0^\varepsilon e^k$ runs over all elements of the centralizer of c_0.

4. Is a Riemann surface symmetric?

We start the section looking for necessary algebraic conditions for a Riemann surface X with a given group G as the group of all its holomorphic automorphisms to be symmetric. Let X and G be represented as $\mathcal{H}/\Gamma$ and Δ/Γ. Let us view Δ as an abstract group δ and let γ be the corresponding normal subgroup of δ isomorphic to Γ. By theorem 2, there is only a finite number of groups $\lambda_1, \ldots, \lambda_n$ which can be realized as NEC groups and which contain δ as a subgroup of index 2. From the proof of this o theorem it follows in addition how such embeddings $\delta \leq \lambda_i$ can look like. Now, this is an entirely algebraic matter to decide which λ_i contain γ as a normal subgroup. Let us call such λ_i, *algebraically admissible for X* and say that λ_i is *conformally admissible* if in addition it can be realized as an NEC group Λ_i containing Δ. It is clear that the existence of algebraically admissible groups is a necessary condition for X to be symmetric.

The converse is not true; the existence of such group λ may not be sufficient for X to be symmetric by reasons of conformal nature. This is because $d(\Lambda) = d(\Lambda^+)/2$ for the Teichmüller dimensions of a proper NEC group Λ and its canonical Fuchsian subgroup Λ^+ [28], which simply saying means that there are "more" Fuchsian groups isomorphic to δ than NEC groups isomorphic to λ. It is worth however to mention that in such a situation a conformal structure on the underlying topological surface can be defined, so that the new surface X' is symmetric, $\text{Aut}^+(X') \cong G$ and the topological characteristics of the action are the same. In fact, any NEC group Λ realizing λ contains groups Δ' and Γ' isomorphic to Δ and Γ respectively and $\Delta'/\Gamma' \cong G$. In the remainder of the lecture we shall see that the conformal structure of X plays a modest game in studies of symmetries, just up to determine G and to decide which of the algebraically admissible groups is actually conformally admissible; the rest depends on G and on the topological characteristics of this action; in one word, on algebra.

The above necessary condition is also sufficient when Δ is a triangle group, say with signature $(0; k, l, m)$ and canonical generators x_1, x_2 and x_3, since then $d(\Delta) = 0$. In such a case any algebraically admissible group is also conformally admissible. Now a Riemann surface X having a group isomorphic to δ/γ as the group of orientation preserving automorphisms corresponds to a pair (a, b) of generators of G of orders k and l respectively and whose product has order m; the surface X can be written as $X = \mathcal{H}/\Gamma$, where $\Gamma = \ker \theta$ for a homomorphism $\theta : \Delta \to G$ defined by the assignment: $\theta(x_1) = a, \theta(x_2) = b$. We shall need the following result proved by Singerman [26].

Theorem 4 *Let X be a Riemann surface corresponding to a generating pair (a, b), where a, b and ab have orders k, l and m respectively and k, l and m are distinct. Then X is symmetric if and only if the mapping $\varphi(a) = a^{-1}, \varphi(b) = b^{-1}$ induces an automorphism of G.*

Proof. We shall prove this theorem in a slightly different way since there is a harmless gap in the original proof. Let $X = \mathcal{H}/\Gamma$ and $G = \Delta/\Gamma = \mathrm{Aut}^+(X)$ for a Fuchsian group Δ with signature $(0; k, l, m)$.

If X is symmetric then there is an NEC group Λ containing Δ as a subgroup of index 2 and Γ as a normal subgroup. Let $\widetilde{G} = \Lambda/\Gamma = \mathrm{Aut}^\pm(X)$. By theorem 2, Λ has signature $(0; +; [-]; \{(k, l, m)\})$. Let c_0, c_1, c_2 be a system of canonical generators of Λ. Then

$$(8) \qquad x_1 = c_0 c_1, \ x_2 = c_1 c_2, \ x_3 = c_2 c_0$$

is a system of canonical generators for Δ. Now

$$(9) \qquad c_1 x_1 c_1 = x_1^{-1} \text{ and } c_1 x_2 c_1 = x_2^{-1}.$$

Thus for $a = \tilde{x}_1$, $b = \tilde{x}_2$, $v = \tilde{c}_1 \in \tilde{\Lambda}/\Gamma$ we have $a^v = a^{-1}$ and $b^v = b^{-1}$ and so $a \mapsto a^{-1}$, $b \mapsto b^{-1}$ induces an automorphism of G indeed. Observe that $\widetilde{G} = \Lambda/\Gamma$ is a semidirect product $G \propto Z_2$.

Conversely suppose that $a \mapsto a^{-1}$, $b \mapsto b^{-1}$ induces an automorphism of G. We have $d(\Delta) = 0$. So there exists an NEC group Λ with signature $(0; +; [-]; \{(k, l, m)\})$ containing Δ. Now a word $w = w(a, b)$ represents identity if and only if $w(x_1, x_2) \in \Gamma$ and $w(a^{-1}, b^{-1}) = 1$ if and only if $w = w(x_1^{-1}, x_2^{-1}) \in \Gamma$. So as $a \mapsto a^{-1}$, $b \mapsto b^{-1}$ induces an automorphism of G we see from (9) that for $w = w(x_1, x_2) \in \Gamma$, $c_1 w(x_1, x_2) c_1 = w(x_1^{-1}, x_2^{-1}) \in \Gamma$ and therefore Γ is normal in Λ; the image σ of c_1 in Λ/Γ can be chosen as a symmetry of X.

5. Quantitative studies of symmetries of Riemann surfaces

The second question stated in section 1 concerns actually the number of real forms of a complex algebraic curve. Given $g \in \mathbb{N}$, let $\nu(g)$ be the maximum number of nonconjugate symmetries with fixed points that a Riemann surface of genus g may admit. The mentioned problem consists in finding the value of the function ν. The first result in this direction is due to Natanzon [23] who proved in 1978 that

$$\nu(g) \leq 2(\sqrt{g} + 1)$$

and that $\nu(g) = 2(\sqrt{g} + 1)$ for each g of the form $g = (2^{n-1} - 1)^2$. Later, in 1994, it was showed in [8] that these are the only values of g for which Natanzon's bound is precise.

At a first view it seems that ν has rather strictly increasing character. But recently were discovered in [9] and [15] that this is not exactly the case; it is true up to certain extent only. What makes this fact surprising is the point that the size and so algebraic complexity of a group of automorphisms G of a Riemann surface X of genus g grows up proportionally, in some sense, to the growing up of g. However, as we shall see, for our purposes, G can be assumed to be a 2-group and under this assumption it turns out that the parity structure of $g - 1$ let to control the algebraic structure of G fairly well. To illustrate it we shall prove with most details the following result from [15].

Theorem 5 *A Riemann surface of even genus g has at most four conjugacy classes of symmetry and this bound is attained for arbitrary even g.*

Proof. At the very beginning notice that when we are looking for the maximal number of conjugacy classes of symmetries of Riemann surfaces it is sufficient to deal with finite 2-groups generated by nonconjugated symmetries. Indeed let X be a Riemann surface of genus g with maximal number of non-conjugate symmetries $\tau_1, \ldots, \tau_m$. Then by the Sylow theorem $\tau_1^{\alpha_1}, \ldots, \tau_m^{\alpha_m}$, generate a 2-subgroup G of $\mathrm{Aut}^{\pm}(X)$ for some $\alpha_1, \ldots, \alpha_m \in \mathrm{Aut}^{\pm}(X)$.

Assume in addition that g is even. Then either G is a cyclic group, a dihedral group, or a semidirect product of a cyclic or dihedral group by Z_2. Indeed let $X = \mathcal{H}/\Gamma$ for some surface Fuchsian group Γ and let $G = \Lambda/\Gamma$. Assume that Λ has a signature of the general form (5). No reflection, nor an elliptic element belong to Γ. Thus G is cyclic or dihedral if Λ has a proper period equal to $|G|$ or a link period equal to $|G|/2$ respectively. So assume that neither a proper period of Λ is equal to $|G|$ nor a link period equal to $|G|/2$. By Riemann-Hurwitz formula

$$g - 1 = \frac{|G|}{2} \left(\alpha g' - 2 + k + \sum_{i=1}^{r} \frac{m_i - 1}{m_i} + \sum_{i=1}^{n} \sum_{j=1}^{s_i} \frac{n_{ij} - 1}{2n_{ij}} \right).$$

So since g is even, we obtain that $m_i = |G|/2$ for some i or $n_{ij} = |G|/4$ for some i, j. In the first case G contains $H = Z_{|G|/2}$ whilst in the second one it contains $H = D_{|G|/4}$ as a subgroup of index 2. Now, in addition G is generated by elements of order 2. Thus there exists an element $z \in G \setminus H$ of order 2 and so $G = H \propto Z_2$. It is easy to show that all semidirect products of a cyclic group by Z_2 have at most three conjugacy classes of elements of order 2. So let $H = \langle x, y \mid x^2, y^2, (xy)^N \rangle$ be dihedral.

In H there exists three conjugacy classes of elements of order 2 represented by x, y and $(xy)^{N/2}$. But only the first two represent symmetries. Now z represents a symmetry of X and each other symmetry from $G \setminus H$ can be written as $z(xy)^\alpha$. Let $\alpha \neq 0$ be as small as possible and let $z(xy)^\beta$ be an element of order 2. We shall show that α divides β. Indeed assume that $\beta = k\alpha + \delta$, where $\delta < \alpha$. Then

$$1 = (z(xy)^\beta)^2 = (z(xy)^\alpha z)^k z(xy)^\delta z(xy)^\delta ((xy)^\alpha)^k$$

which implies that $z(xy)^\delta z(xy)^\delta = 1$ since $z(xy)^\alpha z = (xy)^{-\alpha}$. So $\delta = 0$. But then $z(xy)^\beta = (xy)^{-\alpha k/2} z(xy)^{\alpha k/2}$ or $z(xy)^\beta = (xy)^{-\alpha(k-1)/2} z(xy)^\alpha (xy)^{\alpha(k-1)/2}$ if k is even or odd respectively. This completes the first part of the proof.

In order to prove that for arbitrary even genus g there exists a Riemann surface X having four nonconjugate symmetries consider an NEC group Λ with signature $(0; +; [-]; \{(2, \overset{g+3}{\dots}, 2)\})$. Its Teichmüller space $\mathbf{T}(\Lambda)$ has dimension $2g+3$, by [12] and [27] (cf. [7], p.18) and it is easy to check, using in addition the Hurwitz Riemann formula, that the Teichmüller space of each NEC group Λ_i containing Λ has strictly smaller dimension. As a result the union $\bigcup_{i=1}^n \mathbf{T}(\Lambda_i)$ forms a proper subset of $\mathbf{T}(\Lambda)$ and therefore Λ can be chosen to be maximal i.e, not contained in any other NEC group (cf. Remark 5.1.1(1) and theorem 5.1.2 of [7]). Let $\theta : \Lambda \to G = Z_2 \oplus Z_2 \oplus Z_2 = \langle x, y, z \rangle$ be defined by $\theta(c_{2j}) = x$, $\theta(c_{2j+1}) = y$ for $0 \leq j \leq (g-2)/2$ and

$$\theta(c_i) = \begin{cases} x & \text{for } i = g, g+3, \\ z & \text{for } i = g+1, \\ xyz & \text{for } i = g+2 \end{cases} \quad \text{and} \quad \theta(c_i) = \begin{cases} z & \text{for } i = g, g+2, \\ x & \text{for } i = g+1, g+3 \end{cases}$$

for $g/2$ odd and even respectively. By Lemma 1, $\Gamma = \ker \theta$ is a surface Fuchsian group, and by the Riemann-Hurwitz formula $X = \mathcal{H}/\Gamma$ is a surface of genus g. Finally its symmetries x, y, z and xyz are non-conjugate since $G = \text{Aut}^\pm(X)$. This completes the proof.

Recently, the above result was generalized in [9], where it has been obtained a bound for $\nu(g)$ which depends only on the parity structure of $g - 1$. To be precise, consider the set of all positive integers to be divided in strata, the rth of which N_r consists of all integers of the form $g = 2^{r-1}u + 1$ for some odd u. The above theorem can be understand as the calculation of ν for $g \in N_1$. The main result in [9] can be stated as:

Theorem 6 *Let $g = 2^{r-1}u + 1$, where $r \geq 1$ and u is odd. Then*

$$\nu(g) = \begin{cases} 2^{r+1} & \text{if } u \geq 2^{r+2} - 5, \\ 2^{s-1} & \text{if } (2^{s-1} - 4)/2^{r-s+2} \leq u \leq (2^{s-1} - 4)/2^{r-s+1}, \\ 2^{r-s+2}u + 4 & \text{if } (2^{s-2} - 4)/2^{r-s+2} \leq u \leq (2^{s-1} - 4)/2^{r-s+2}, \end{cases}$$

where $s \leq r+2$ is the unique integer for which u is in range $(2^{s-2}-4)/2^{r-s+2} \leq u \leq (2^{s-1} - 4)/2^{r-s+1}$.

6. Qualitative studies of symmetries of Riemann surfaces

Recall ([11]) that the topological type of a symmetry σ of a Riemann surface X is determined by a triple (g, k, ε), where g is the genus of X, k the number of connected components of $\text{Fix}(\sigma)$, to which we shall refer as to the set of *ovals* of σ, and $\varepsilon = 1$ or $\varepsilon = 0$ according to whether $X \setminus \text{Fix}(\sigma)$ is connected or not. This explains why σ is said to be *separating* for $\varepsilon = 0$. Let us call such triple the *topological type of the symmetry σ*. Distinct types of symmetries of torus are illustrated in Figure 5. We start this section showing the classical result of Harnack.

Theorem 7 *A triple (g, k, ε) is the topological type of some symmetry if and only if*
(10) $$1 \leq k + \varepsilon \leq g + 1 \text{ and } g + 1 \equiv k \bmod (2 - \varepsilon).$$

Proof. Let σ be a symmetry of a Riemann surface X of genus $g \geq 2$. Then $X = \mathcal{H}/\Gamma$ for some Fuchsian surface group Γ and $\langle \sigma \rangle = \Lambda/\Gamma$. Now, since $\Lambda^+ = \Gamma$, Λ has signature $(g'; \pm, [-]; \{(-), .^k., (-)\})$, where $k = g + 1 - \delta g'$, by theorem 2. In addition $\varepsilon = 0$ means that Λ has the sign $+$, i.e, $\delta = 2$.

Conversely assume that (g, k, ε) satisfies (10). Let first $\varepsilon = 0$. Consider an NEC group Λ with signature $(g'; +; [-]; \{(-), .^k., (-)\})$, where $g' = (g + 1 - k)/2$ and an epimorphism $\theta : \Lambda \to Z_2 = \langle a \rangle$ given by $\theta(c_i) = a$, $\theta(a_i) = \theta(b_i) = \theta(e_i) = 1$. Then, by Lemma 1, $\Gamma = \ker \theta$ is a Fuchsian surface group of orbit genus g. Thus $X = \mathcal{H}/\Gamma$ is a Riemann surface of genus g having a symmetry of type $(g, k, 0)$.

For $\varepsilon = 1$ let $g' = g + 1 - k$ and let Λ be an NEC group with signature $(g'; -; [-]; \{(-), .^k., (-)\})$. Let $\theta : \Lambda \to Z_2 = \langle a \rangle$ be an epimorphism given by $\theta(c_i) = a$, $\theta(d_i) = a$, $\theta(e_i) = 1$. Then again by Lemma 1, $\Gamma = \ker \theta$ is a Fuchsian surface group of orbit genus g and therefore $X = \mathcal{H}/\Gamma$ is a Riemann surface of genus g having a symmetry of type $(g, k, 1)$. This completes the proof.

It is convenient to denote the topological type (g, k, ε) of a symmetry of a Riemann surface of genus g by $+k$ if $\varepsilon = 0$ and $-k$ otherwise. This integer is called usually *the species* $\text{sp}(\sigma)$ of σ. The above theorem gives necessary and

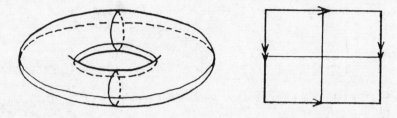

Two separating symmetries corresponding to reflections of fundamental region (on the right); each of them has two ovals; a sphere with two holes as orbit space.

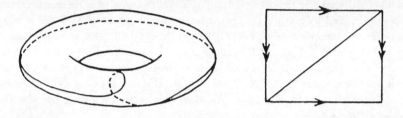

A nonseparating symmetry corresponding to a reflection with respect to a diagonal of a fundamental region; it has one oval; a projective plane with hole as the orbit space.

A symmetry without fixed points corresponding to a glide reflection of the plane; a Klein bottle as the orbit space.

Figure 5

sufficient conditions for an integer to be the species of a symmetry of some Riemann surface of genus g. Call the unordered s-uple $[\mathrm{sp}(\sigma_1), \ldots, \mathrm{sp}(\sigma_s)]$, for representatives $\{\sigma_1, \ldots, \sigma_s\}$, of all conjugacy classes of symmetries of a Riemann surface X of genus g to be the *symmetry type of* X. A lot of attention has been payed to the problem of finding necessary and sufficient conditions for an unordered s-uple of integers $[k_1, \ldots, k_s]$ to be the symmetry type of some Riemann surface of genus g. However a complete answer to this question is known only for $s = 2$. One can however look for some necessary condition in the general case, or try to find necessary and sufficient conditions for certain particular classes of surfaces; complete solutions are known for example for surfaces having $PSL(2, q)$ as Hurwitz automorphism groups [3], for the so called Accola - Maclachlan - Kulkarni surfaces [4] and for hyperelliptic Riemann surfaces [10]. Here we shall concentrate ourselves on some aspects concerning the second of these classes, already discussed in [11].

In section 3 we have mentioned that in studies of symmetries, the conformal structure of a Riemann surface plays a game just up to certain extend. We already could observe it in the previous section. Here we shall illustrate it in the context of qualitative studies. First we shall show how to find the number of ovals of a symmetry σ of Riemann surface $X = \mathcal{H}/\Gamma$ from $G = \mathrm{Aut}^{\pm}(X)$ and from the topological characteristics of this action. Let $G = \Lambda/\Gamma$ and let $\theta : \Lambda \to G$ be the canonical projection. Then $\langle \sigma \rangle = \Gamma_\sigma/\Gamma$ for some NEC subgroup Γ_σ of Λ and from the one hand $\mathrm{Fix}(\sigma)$ is homeomorphic to the boundary of X/σ, whilst on the other hand it is the number of empty period cycles of Γ_σ since $X/\sigma \cong \mathcal{H}/\Gamma_\sigma$. In our case groups Γ_σ are usually non normal subgroups of Λ but an advantage of our situation is that we know a priori that they are surface NEC groups since $\Gamma_\sigma^+ = \Gamma$. Following [13] we shall present now a method of finding the number of period cycles of Γ_σ mentioned at the beginning of section 3.

For, we have to count reflections of Λ which are in Γ_σ and which are not conjugated there. Observe that $\Gamma_\sigma = \theta^{-1}(\langle \sigma \rangle)$. So if σ has fixed points then it is conjugate to $\theta(c_i)$ for some canonical reflection c_i of Λ; without loss of generality we can assume that $\theta(c_i) = \sigma$ since conjugate symmetries have the same number of ovals. Now given $w \in \Lambda$, $c_i^w \in \Gamma_\sigma$ if and only if $w \in \theta^{-1}(C(G, \theta(c_i)))$, the inverse image of the centralizer of $\theta(c_i)$ in G which we shall denote by C_i. In particular we see that C_i normalizes Γ_σ and thus for $v, w \in C_i$, the reflections c_i^v and c_i^w of Γ_σ are conjugate in Γ_σ if and only if $w^{-1}v \in C(\Lambda, c_i)\Gamma_\sigma$. As a consequence, the conjugates of c_i give rise to

$$[C_i : C(\Lambda, c_i)\Gamma_\sigma] = [C(G, \sigma) : \theta(C(\Lambda, c_i))]$$

empty period cycles in Γ_σ.

Let now $c_j^w \in \Gamma_\sigma$ for some $j \neq i$ and $w \in \Lambda$. Then $wC_jw^{-1} = C_i$. Furthermore $c_j^v \in \Gamma_\sigma$ if and only if $vw^{-1} \in C_i$. Finally given $u, u' \in C_i$ and $v = uw, v' = u'w$, the reflections $c_j^v, c_j^{v'}$ are conjugate in Γ_σ if and only if $v^{-1}v' \in C(\Lambda, c_j)w^{-1}\Gamma_\sigma w = C(\Lambda, c_j)\Gamma$ which means that $u^{-1}u' \in wC(\Lambda, c_j)\Gamma w^{-1}$. So the conjugates of c_j give rise to

$$[C_i : wC(\Lambda, c_j)\Gamma w^{-1}] = [C_j : C(\Lambda, c_j)\Gamma] = [C(\theta(\Lambda), \theta(c_j)) : \theta(C(\Lambda, c_j))]$$

empty period cycles in Γ_σ. Summing up we arrived to the following

Theorem 8 *A symmetry σ with fixed points of a Riemann surface X is conjugate to $\theta(c_i)$ for some canonical reflection c_i of Λ and it has*

$$\sum [C(G, \theta(c_j)) : \theta(C(\Lambda, c_j))]$$

ovals, where c_j is running over nonconjugate canonical reflections of Λ, whose images under θ belongs to the orbit of σ in G.

As an application we shall show how to find the number of ovals for symmetries of Accola - Maclachlan - Kulkarni surfaces.

Given an integer $g \geq 2$ let $N(g)$ be the order of the largest group of conformal automorphism that a compact Riemann surface admits. Hurwitz [18] showed that $N(g) \leq 84(g - 1)$ and Macbeath [21] that this bound can not be sharpened for infinitely many values of g. In the opposite direction, Accola [1] and Maclachlan [22] independently showed that $N(g) \geq 8(g + 1)$ and for each g the surface X_g given by $y = x^{2(g+1)} - 1$ has a group of automorphisms of order $8(g + 1)$. They also showed that a group of automorphisms of bigger order can not be constructed for infinitely many g. Later on, Kulkarni showed, that for large values of g, $g \not\equiv 3 \bmod 4$, there exists only one surface X_g whose group of automorphisms has order $8(g + 1)$ whilst for $g \equiv 3 \bmod 4$ there exists exactly one more surface Y_g whose group of automorphisms has such order and that

$$G = \mathrm{Aut}^+(X_g) = \langle a, b \mid a^{2g+2}, b^4, (ab)^2, ab^2 a^{-1} b^2 \rangle,$$
$$G = \mathrm{Aut}^+(Y_g) = \langle a, b \mid a^{2g+2}, b^4, (ab)^2, b^2 ab^2 a^g \rangle.$$

Let Δ be the Fuchsian group with signature $(0; 2g + 2, 4, 2)$. These surfaces are constructed as $\mathcal{H}/\Gamma$, where Γ is the kernel of the epimorphism $\Delta \to G$ defined by assignment: $x_1 \mapsto a, x_2 \mapsto b$. Clearly $a \mapsto a^{-1}, b \mapsto b^{-1}$ induces an automorphism of the above groups. So X_g and Y_g are symmetric, by theorem 4. Now every NEC group Λ containing Δ as the cannonical Fuchsian subgroup has signature $(0; +; [-]; \{(2g + 2, 4, 2)\})$ and is unique up to conjugation in $\mathcal{G}$. So Γ is a normal subgroup of Λ and therefore from the proof of theorem 4 it follows that $\mathrm{Aut}^{\pm}(X_g)$ and $\mathrm{Aut}^{\pm}(Y_g)$ have the presentations

(11) $\quad \langle x, y, z \mid x^2, y^2, z^2, (xy)^{2g+2}, (yz)^4, (xz)^2, (x(yz)^2)^2 \rangle$ and

(12) $\quad \langle x, y, z \mid x^2, y^2, z^2, (xy)^{2g+2}, (yz)^4, (xz)^2, z(xy)^2 z(xy)^{g-1} \rangle$

respectively. The canonical epimorphism $\theta : \Lambda \to \Lambda/\Gamma$ is defined by the assignment $\theta(c_0) = x, \theta(c_1) = y, \theta(c_2) = z$. We are looking for conjugacy classes of elements of order 2 which represent orientation reversing automorphisms. Observe that $(yz)^2$ generates a central subgroup H of order 2 in $G = \mathrm{Aut}^{\pm}(X_g)$ and $G/H = D_{2g+2} \oplus Z_2 = \langle \tilde{x}, \tilde{y} \rangle \times \langle \tilde{z} \rangle$. So each element of G can be written in the form $(xy)^k x^\alpha z^\beta (yz)^{2\gamma}$ for unique $0 \leq k \leq 2g + 1, 0 \leq \alpha, \beta, \gamma \leq 1$. Now such element has order 2 and represents an orientation reversing automorphism if and only if $(xy)^k x^\alpha z^\beta$ has order 2 and odd length. If $\beta = 0$ then there are two conjugacy classes of such elements: $[x]$ and $[y]$. If $\beta = 1$, then $\alpha = 0$. But $1 = (xy)^k z (xy)^k z = (xy)^k (xzyz)^k = (xy)^{2k} (yz)^{2k}$. So $(xy)^k z$ represents an element of order 2 if and only if $k = 0$ or else $k = g + 1$ if g is odd. Thus among elements of order 2 of the form $(xy)^2 x^\alpha z^\beta$ we have 3 conjugacy classes represented by x, y, z and one more class represented by $(xy)^{g+1} z$ if g is odd. Finally we have to look for conjugacy classes of elements of order 2 of the form $(xy)^k x^\alpha z^\beta (yz)^2$. Since $(yz)^2$ is a central element of G, two such elements are conjugate if and only if their "parts" $(xy)^k x^\alpha z^\beta$ are conjugate. But $y(yz)^2 \sim y$, $z(yz)^2 \sim z$. So still remains to consider the class $[x(yz)^2]$ and the class $[(xy)^{g+1} z (yz)^2]$ if g is odd. However $(xy)^{g+1} z (yz)^2 \sim (xy)^{g+1} z$ since

$$(xy)^{g+1} z (yz)^2 = (xy)^g x (yz)^3 = (xy)^{-1} (xy)^{g+1} zxy.$$

Clearly $x(yz)^2$ is not conjugate to y, z or $(xy)^{g+1} z$ since their images in $\widetilde{G}$ are not conjugate. Finally we have

(13) $$(xy)^k x^\alpha z^\beta (yz)^{2\gamma} x ((xy)^k x^\alpha z^\beta (yz)^{2\gamma})^{-1} = (xy)^{2k} x$$

for arbitrary α, β, γ, k since $(yz)^2$ is central and z commutes with x. So $x \not\sim x(yz)^2$. Summing up, we see that X_g has three conjugacy classes of symmetries with fixed curves represented by x, y and z, one without fixed points represented by $x(yz)^2$ and one more without fixed points represented by $(xy)^{g+1} z$ if in addition g is odd.

Finally we shall determine the number of ovals for x for example. Equation (13) shows that the orbit of x has $g + 1$ elements. In particular $|C(G, x)| = 16$ since $|G| = 16(g + 1)$. In addition $\theta((c_0 c_1)^{g+1} (c_0 c_2)) = (xy)^{g+1} xz$ has order 2 if g is odd and 4 if g is even. So, by theorem 3, $\theta(C(\Lambda, c_0))$ has order 8 and 16 respectively. Therefore, by theorem 8 we see that x represents a symmetry with 2 ovals if g is odd and 1 oval otherwise.

Theorem of Harnack from the beginning of this section says in particular that a single symmetry of a Riemann surface of genus $g \geq 2$ has at most $g + 1$ ovals. Much later S. M. Natanzon [24] (cf. also [25]) proved that k in range $2 \leq k \leq 4$ symmetries of a Riemann surface of genus g admits at most $2g + 2^{k-1}$ ovals in total and that this bound is attained respectively for every g congruent to 1 modulo 2^{k-2}. A Riemann surface of even genus g has at most four nonconjugate symmetries and the total number of their ovals has been found in [16]. This problem was also studied by Singerman [29], who showed that for arbitrary k there exist infinitely many values of g for which there exists a Riemann surface of genus g having k non-conjugate symmetries, $M_k = 2g + 2^{k-3}(9 - k) - 2$ ovals in total and conjectured that this is a bound for the total number of ovals of k symmetries of a Riemann surface of genus g. The last turns out to be false for $k > 9$; we finish our lecture stating without proofs a result from [14], which can be proved using the methods described above.

Theorem 9 *Arbitrary $k, k \geq 9$ non-conjugate symmetries of a Riemann surface of genus $g \geq 2$ have at most $2g - 2 + 2^{r-3}(9 - k)$, ovals, where r is the smallest positive integer for which $k \leq 2^{r-1}$.*

Furthermore the above bound is sharp for each $k \geq 9$ for infinitely many g. Namely we have

Theorem 10 *Let $k \geq 9$ be an arbitrary integer and let r be the smallest positive integer for which $k \leq 2^{r-1}$. Then for arbitrary $g = 2^{r-2}t + 1$, where $t \geq k - 3$ there exists a Riemann surface X of genus g having k non-conjugate symmetries which have $2g - 2 + 2^{r-3}(9 - k)$ ovals in total.*

References

[1] Accola R. D. M.: On the number of automorphisms of a closed Riemann surface. Trans. Amer. Math. Soc. **131** (1968), 398–408.

[2] Alling N. L., Greenleaf N.: *Foundations of the theory of Klein surfaces.* Lect. Notes in Math. **219**, Springer-Verlag (1971).

[3] Broughton A., Bujalance E., Costa A. F., Gamboa J. M., Gromadzki G.: Symmetries of Riemann surfaces on which PSL$(2, q)$ acts as Hurwitz automorphisms group, Journal Pure Appl. Algebra **106**(2) (1996) 113-126.

[4] Broughton A.F., Bujalance E., Costa A.F., Gamboa J.M., Gromadzki G.: Symmetries of Accola–Maclachlan–Kulkarni surfaces, Proc. Amer. Math. Soc. **127**(3) (1999), 637-646.

[5] Bujalance E.: Normal subgroups of NEC groups. Math. Z. **178** (1981), 331–341.

[6] Bujalance E.: Proper periods of normal NEC subgroups with even index. Rev. Mat. Hisp.-Amer. (4), **41** (1981), 121–127.

[7] Bujalance E., Etayo J. J., Gamboa J. M., Gromadzki G.: *A combinatorial approach to groups of automorphisms of bordered Klein surfaces,* Lecture Notes in Math. vol. **1439**, Springer Verlag (1990).

[8] Bujalance E., Gromadzki G., Singerman D.: On the number of real curves associated to a complex algebraic curve, Proc. Amer. Math. Soc. **120**(2) (1994) 507-513.

[9] Bujalance E., Gromadzki G., Izquierdo M., On Real Forms of a Complex Algebraic Curve, to appear in Journal Australian Math. Soc.

[10] Bujalance E., Cirre J., Gamboa J.M., Gromadzki G.: Symmetry Types of Hyperelliptic Riemann Surfaces, to appear.

[11] Cirre J., Gamboa J.M.: Compact Klein surfaces and real algebraic curves, this volume.

[12] Fricke R., Klein F.: *Vorlesungen über die Theorie der automorphen Funktionen.* (2 vols.). B. G. Teubner, Leipzig (1897 and 1912).

[13] Gromadzki G.: On a Harnack-Natanzon theorem for the family of real forms of Riemann surfaces, Journal Pure Appl. Algebra **121** (1997), 253-269.

[14] Gromadzki G.: On ovals of Riemann surfaces, to appear in Rev. Matemática Iberoamericana.

[15] Gromadzki G., Izquierdo M.: Real forms of a Riemann surface of even genus, Proc. Amer. Math. Soc. **126** (12) 1998, 3475-3479.

[16] Gromadzki G., Izquierdo M.: On ovals of Riemann surfaces of even genera (1998), Geometriae Dedicata **78** (1999), 81-88.

[17] Hoare A. H. M.: Subgroups of NEC groups and finite permutation groups. Quart J. Math. Oxford (2), **41** (1990), 45-59.

[18] Hurwitz A.: Über algebraische Gebilde mit eindeutigen Transformationen in sich. Math. Ann. **41** (1893), 402–442.

[19] Kulkarni R. S.: A note on Wiman and Accola-Maclachlan surfaces. Ann. Acad. Sci. Fenn. **16** (1991), 83–94.

[20] Macbeath A. M.: The classification of non-euclidean crystallographic groups. Can. J. Math. **19** (1967), 1192–1205.

[21] Macbeath A. M.: On a theorem of Hurwitz. Proc. Glasgow Math. Assoc. **5** (1961), 90 96.

[22] Maclachlan C.: A bound for the number of automorphisms of a compact Riemann surface. J. London Math. Soc. **44** (1969), 265–272.

[23] Natanzon S. M.: On the order of a finite group of homeomorphisms of a surface into itself and the real number of real forms of a complex algebraic curve. Dokl. Akad. Nauk SSSR **242** (1978), 765–768. (Soviet Math. Dokl. **19** (5), (1978), 1195–1199.)

[24] Natanzon S. M.: On the total number of ovals of real forms of complex algebraic curves. Uspekhi Mat. Nauk **35**:1 (1980), 207–208. (Russian Math. Surveys **35**:1 (1980), 223–224.)

[25] Natanzon S. M.: Finite group of homeomorphisms of surfaces and real forms of complex algebraic curves, Trudy Moskov. Mat. Obshch. **51** (1988) (Trans. Moscow Math. Soc. **51** (1980),1-51.)

[26] Singerman D.: Non-Euclidean crystallographic groups and Riemann surfaces. Ph. D. Thesis, Univ. of Birmingham (1969).

[27] Singerman D.: Symmetries of Riemann surfaces with large automorphism group. Math. Ann. **210** (1974), 17–32.

[28] Singerman D.: On the structure of non-euclidean crystallographic groups. Proc. Camb. Phil. Soc. **76** (1974), 233–240.

[29] Singerman D.: Mirrors on Riemann Surfaces, Contemporary Mathematics, **184** (1995), 411-417.

[30] Wilkie H. C.: On non-euclidean crystallographic groups. Math. Z. **91** (1966), 87–102.

Institute of Mathematics
University of Gdańsk
Wita Stwosza 57
80-952 Gdańsk Poland
greggrom@math.univ.gda.pl

COMPACT KLEIN SURFACES
AND REAL ALGEBRAIC CURVES

F. J. Cirre and J. M. Gamboa [1]

A classical Riemann surface is a topological surface without boundary together with an analytic structure. This analytic structure makes it orientable. However, nonorientable or orientable surfaces with boundary may admit a *dianalytic structure* which behaves in many aspects as the analytic structure of a classical Riemann surface. Roughly speaking, a dianalytic structure is the equivalence class of an atlas whose transition functions are either analytic or antianalytic (a function is antianalytic if its composite with complex conjugation is analytic). For a rigorous definition of dianalytic structure see [2]. The easiest way in which these non-classical surfaces arise is as quotients of classical Riemann surfaces under the action of *antianalytic* involutions.

For example, the surface with non-empty boundary K in the figure is the quotient of the Riemann surface X under the antianalytic involution $\sigma : X \to X$ induced by the symmetry with respect to the plane containing the boundary of K.

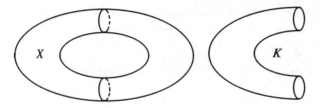

Such a topological space K is endowed with a dianalytic structure with respect to which the projection

$$\pi : X \to X/\sigma = K$$

"behaves well", *i.e.*, it is a morphism in the category of surfaces with dianalytic structures. Note that if X is compact then K is compact too.

Objects such as K are usually known as Klein surfaces. An equivalent definition, which is the one we shall use here, is the following: a *Klein surface* is a pair (X, σ) where X is a Riemann surface and $\sigma : X \to X$ is an antianalytic

[1]Both authors partially supported by DGICYT PB98-0756

involution on it. Some authors call these pairs (X, σ) *symmetric Riemann surfaces.*

A *morphism between two Klein surfaces* (X, σ) and (Y, τ) is an analytic map

$$\varphi : X \to Y$$

which is "compatible" with σ and τ, *i.e.*, such that

$$\varphi \circ \sigma = \tau \circ \varphi.$$

In particular, (X, σ) and (Y, τ) are *isomorphic* if there exists an isomorphism of Riemann surfaces $\varphi : X \to Y$ with $\varphi \circ \sigma = \tau \circ \varphi$. In such a case we shall write $(X, \sigma) \equiv (Y, \tau)$. This is the classical approach used by pioneers in the study of the real theory of algebraic curves during the last quarter of the past century. Among them, the works of Gordon, Harnack, Hurwitz, Klein, and Weichold stand out.

The goal of this note is to present the problem of classification of compact Klein surfaces up to isomorphism. For that, we discuss factors which obstruct two Klein surfaces from being isomorphic, and we analyze some examples. We work in the algebraic category of smooth real algebraic curves, which is equivalent to that of compact Klein surfaces. This was first explained by Alling and Greenleaf in [2]. We present here a brief description of this equivalence adapted to our definition of a Klein surface.

The algebraic object that completely determines the compact Klein surface (X, σ) is its field $\mathcal{M}(X, \sigma)$ of meromorphic functions on (X, σ). To introduce this object we endow the Riemann sphere $\widehat{\mathbb{C}} = \mathbb{C} \cup \{\infty\}$ with the structure of a Klein surface by fixing the antianalytic involution

$$c : \widehat{\mathbb{C}} \to \widehat{\mathbb{C}} : z \mapsto \begin{cases} \overline{z} & \text{if } z \in \mathbb{C}, \\ \infty & \text{otherwise.} \end{cases}$$

This way a meromorphic function on (X, σ) is nothing but a morphism between (X, σ) and $(\widehat{\mathbb{C}}, c)$, *i.e.*, a meromorphic function $f \in \mathcal{M}(X)$ on the Riemann surface X such that $c \circ f = f \circ \sigma$. In particular, each real number can be identified as a constant meromorphic function on (X, σ), but $\sqrt{-1} \notin \mathcal{M}(X, \sigma)$. It is easily seen that $\mathcal{M}(X) = \mathcal{M}(X, \sigma)(\sqrt{-1})$. Hence $\mathcal{M}(X, \sigma)$ *is an algebraic function field in one variable over the field* $\mathbb{R}$ *of real numbers* as $\mathcal{M}(X)$ is so over $\mathbb{C}$ (see, for example, the article of Turbek [22] in this volume).

Conversely, to each algebraic function field $E \,|\, \mathbb{R}$ in one variable over $\mathbb{R}$ such that $\sqrt{-1} \notin E$, we can associate a compact Klein surface (X, σ) whose field of meromorphic functions is $\mathbb{R}$-isomorphic to E. Let us explain this. Consider the field $F = E(\sqrt{-1})$, which is an algebraic function field in one variable over $\mathbb{C}$. Recall that a valuation ring of F (over $\mathbb{C}$) is a subring V of F containing $\mathbb{C}$ such that $1/f \in V$ for each $f \in F - V$. Each valuation ring V is a local ring with maximal ideal $m_V = \{f \in V^* : 1/f \notin V\} \cup \{0\}$ and whose residue field $k(V) = V/m_V$ is isomorphic to $\mathbb{C}$ via the canonical inclusion $\mathbb{C} \hookrightarrow k(V) : a \mapsto a + m_V$, see [10]. Analogously, a valuation ring of a field extension $E \mid \mathbb{R}$ is a subring W of E containing $\mathbb{R}$ such that $1/f \in W$ for each $f \in F - W$. In this case the residue field $k(W) = W/m_W$ is isomorphic to either $\mathbb{C}$ or $\mathbb{R}$.

Each $f \in F$ can be viewed as a function on the set $X(F)$ of all valuation rings of F defined as

$$ f : X(F) \to \widehat{\mathbb{C}} : V \mapsto \begin{cases} f + m_V & \text{if } f \in V, \\ \infty & \text{otherwise.} \end{cases} $$

In this way, $X(F)$ is endowed with the initial topology with respect to the family of functions $\{f : X(F) \to \widehat{\mathbb{C}} : f \in F\}$. With respect to this topology, $X(F)$ is a compact space, since the map

$$ X(F) \to \prod_{f \in F} \widehat{\mathbb{C}} : V \mapsto (f(V) : f \in F) $$

is a closed topological embedding. In fact, $X(F)$ admits a structure of a compact Riemann surface by using the functions induced by the generators of the maximal ideals m_V, for $V \in X(F)$, as local coordinates. Moreover, the meromorphic functions on $X(F)$ are precisely the functions $f : X(F) \to \widehat{\mathbb{C}}$ induced by elements $f \in F$ (see [10] or [22]). In other words, $\mathcal{M}(X(F)) = F$.

In the case we are dealing with, $F = E(\sqrt{-1})$ and so each element in F can be written in a unique way as $a + b\sqrt{-1}$ with $a, b \in E$. So, the assignment $\sigma(F) : X(F) \to X(F) : V \mapsto \sigma(F)(V)$ where

$$ \sigma(F)(V) = \{a - b\sqrt{-1} : a + b\sqrt{-1} \in V\} $$

is a well defined mapping. Moreover, it is an antianalytic involution on the compact Riemann surface $X(F)$, see [13]. Therefore *associated to each algebraic function field in one variable over $\mathbb{R}$ there is a compact Klein surface.*

Moreover, the meromorphic functions of such a compact Klein surface $(X(F), \sigma(F))$ are the functions $f : X(F) \to \widehat{\mathbb{C}}$ induced by the elements $f \in E$. In other words, *the field of meromorphic functions of the Klein surface associated to an algebraic function field E in one variable over $\mathbb{R}$ is (isomorphic to) E.* Conversely, if (X, σ) is a compact Klein surface and $E := \mathcal{M}(X, \sigma)$, then (X, σ) *is isomorphic to* $(X(F), \sigma(F))$, where $F = E(\sqrt{-1})$. Indeed, it is easily seen [13] that the isomorphism of compact Riemann surfaces

$$ H : X \to X(F) : p \mapsto \{f \in \mathcal{M}(X) : f \text{ is holomorphic at } p\} $$

satisfies the equality $\sigma(F) \circ H = H \circ \sigma$ and so it is an isomorphism between the compact Klein surfaces (X, σ) and $(X(F), \sigma(F))$.

In this way the equivalence between the categories of compact Klein surfaces and algebraic function fields in one variable over $\mathbb{R}$ is established. Let us go deeper into this equivalence.

The *real part* of the Klein surface (X, σ) is defined as the fixed point set of σ:

$$X_\sigma = \{p \in X : \sigma(p) = p\}.$$

The reason for this name is clear if we deal with $(X(F), \sigma(F))$ instead of (X, σ). Note that for each $V \in X(F)$, the intersection $W = V \cap E$ is a valuation ring of E (over $\mathbb{R}$), and the real part of $(X(F), \sigma(F))$ is

$$X(F)_{\sigma(F)} = \{V \in X(F) : k(W) = \mathbb{R}\}.$$

Indeed, suppose that $V \in X(F)$ is fixed by $\sigma(F)$ but $k(W) \neq \mathbb{R}$. Then $\sqrt{-1} \in k(W)$, *i.e.*, there exists $f \in W \subset V$ such that $-1 = f^2 + m_W$. Thus $(f + \sqrt{-1})(f - \sqrt{-1}) = f^2 + 1 \in m_W \subset m_V$ and m_V being a prime ideal, we can assume without loss of generality that $f + \sqrt{-1} \in m_V$. Hence, $f - \sqrt{-1} \in m_{\sigma(F)(V)}$, and substracting, $2\sqrt{-1} \in m_V$, a contradiction. Conversely, suppose that $\sigma(F)(V) \neq V$. Then, there exist $f, g \in F$ such that $f + \sqrt{-1}g \in V$ but $f - \sqrt{-1}g \notin V$. We can assume without loss of generality that $h = f/g \in W$. Obviously neither f nor g belongs to W and so $1/g \in m_V$. But $1 + h^2 = (f + \sqrt{-1}g)(h - \sqrt{-1})/g$, which belongs to $m_V \cap E = m_W$, and so $\sqrt{-1} \in k(W) = \mathbb{R}$, a contradiction.

An important example of how a compact Klein surface is determined by its field of meromorphic functions, is the translation into algebraic language of whether its real part is empty or not. Recall that a field K admits an *ordering* (compatible with sum and product) if and only if K is *real*, *i.e.*, -1 is not a sum of squares of elements in K. We can prove the following.

Theorem 1. *The real part of the compact Klein surface (X, σ) is nonempty if and only if its field of meromorphic functions $\mathcal{M}(X, \sigma)$ is real.*

Proof: We can substitute (X, σ) by $(X(F), \sigma(F))$ where $F = E(\sqrt{-1})$ and $E = \mathcal{M}(X, \sigma)$. Assume first that E is real. Let $\leq$ be an ordering in E and let us consider the following valuation ring of E:

$$W = \{f \in E : -r \leq f \leq r \text{ for some } r \in \mathbb{R}\}.$$

Clearly the ring $V = \{f + \sqrt{-1}g : f, g \in W\}$ is a valuation ring of F, *i.e.*, $V \in X(F)$, such that $V \cap E = W$. So, to prove that (X, σ) has non-empty real part it suffices to check that $k(W) = \mathbb{R}$. Were it not so, then there would exist some $f \in W$ such that $1 + f^2 \in m_W$. Hence $x = -(1 + f^2) \in m_W \subset m_V$ and $0 < 1 + x$ since otherwise $0 \leq -1/x \leq 1$ and in particular $-1/x \in V$, a contradiction because $x \in m_V$. Thus $0 < 1 + x = -f^2$, absurd.

Conversely, assume now that (X, σ) has nonempty real part and take a valuation ring $V \in X(F)$ such that $k(W) = \mathbb{R}$ for $W = V \cap E$. To prove that

E is a real field, assume by the way of contradiction that $-1 = f_1^2 + \cdots + f_n^2$ for some $f_i \in E$. Let us write $f_0 = 1$. Now, given $n+1$ elements $f_0, f_1, \ldots, f_n$ in $E - \{0\}$ it is easy to prove by induction on n that there exists $i_0 \in \{0, \ldots, n\}$ such that $g_j := f_j/f_{i_0}$ belongs to W for all $j \in \{0, \ldots, n\}$. In our case, this gives $-1 = \sum_{j \neq i_0} g_j^2$. Then

$$-1 + m_W = \sum_{j \neq i_0} (g_j + m_W)^2,$$

that is, -1 is a sum of squares in $k(W) = \mathbb{R}$, absurd.

Example 1. We now present a nonorientable compact Klein surface (X, σ) with empty real part. (A Klein surface (X, σ) is nonorientable if $X/\langle \sigma \rangle$ is.) Consider the irreducible polynomial $f(x, y) = x^2 + y^2 + 1$ and the quotient field E of the domain $\mathbb{R}[x, y]/(f)$. It is an algebraic function field in one variable over $\mathbb{R}$ and $\sqrt{-1} \notin E$. Otherwise, there would exist polynomials $g, h \in \mathbb{R}[x, y]$ such that $g^2 + h^2 \in (f)$ and $h \notin (f)$, i.e.,

$$(g + \sqrt{-1}h)(g - \sqrt{-1}h) \in f \cdot \mathbb{C}[x, y].$$

Since f is also irreducible as a polynomial in $\mathbb{C}[x, y]$ we can suppose that

$$g + \sqrt{-1}h = (a + \sqrt{-1}b)f, \quad \text{with } a, b \in \mathbb{R}[x, y]$$

and so $h = bf \in (f)$, which is false. Thus $\sqrt{-1} \notin E$ and we can repeat the above construction: considering $F = E(\sqrt{-1})$ it turns out that the compact Riemann surface $X(F)$ admits an antianalytic involution $\sigma(F)$. Therefore, $(X, \sigma) := (X(F), \sigma(F))$ is a compact Klein surface and, according to the above theorem, its real part is empty. Indeed, its field E of meromorphic functions is not real because

$$-1 = (x + (f))^2 + (y + (f))^2.$$

Now the quotient $X/\langle \sigma \rangle$ of a compact Riemann surface under an antianalytic involution either has non-empty boundary or is non-orientable (or both). In our case the first possibility does not occur and so the compact Klein surface (X, σ) is nonorientable. Further, it is not difficult to see that as a topological surface $X/\langle \sigma \rangle$ has genus 1 and so it is homeomorphic to the real projective plane, see [13].

Algebraic function fields in one variable over $\mathbb{R}$ not containing $\sqrt{-1}$ correspond bijectively with *smooth real algebraic curves* (up to isomorphism). This is the most convenient way to present compact Klein surfaces for our purposes. Firstly we must introduce this notion precisely.

We begin by recalling the notion of a *complex* curve. Let us denote by $\mathbb{P}^n(\mathbb{C})$ the n-dimensional projective space over the field $\mathbb{C}$ of complex numbers. A *smooth complex projective curve* is a set

$$(1) \qquad X = \{x = [x_0 : \cdots : x_n] \in \mathbb{P}^n(\mathbb{C}) : F_1(x) = \cdots = F_m(x) = 0\},$$

where $F_1, \ldots, F_m \in \mathbb{C}[x_0, \ldots, x_n]$ are homogeneous polynomials such that for each point $p \in X$ there exists a finite subset I_p of $\{1, \ldots, m\}$ with $\text{card}(I_p) = n - 1$ satisfying two conditions:

i) for some open neighbourhood U of p in $\mathbb{P}^n(\mathbb{C})$,

$$X \cap U = \{x \in \mathbb{P}^n(\mathbb{C}) : F_i(x) = 0, \text{ for all } i \in I_p\},$$

ii) the rank of the matrix

$$\left(\frac{\partial F_i}{\partial x_j}(p)\right)_{i \in I_p,\ 0 \le j \le n}$$

is maximum, *i.e.*, $n - 1$.

An immediate consequence of the Implicit Function Theorem is that X admits a structure of compact Riemann surface.

One of the main consequences of the Riemann-Roch theorem is the converse of this result: every compact Riemann surface is analytically embedded in some projective space $\mathbb{P}^n(\mathbb{C})$. More precisely:

Theorem 2. *Let X be a compact Riemann surface of genus g. Then there exist meromorphic functions $f_0, \ldots, f_{g+1}$ on X such that the map*

$$\Phi : X \to \mathbb{P}^{g+1} : x \mapsto [f_0(x) : \cdots : f_{g+1}(x)]$$

is an isomorphism between X and its image $\Phi(X)$, which is a smooth complex projective curve.

Assume now that X is a smooth complex projective curve defined as in (1), such that the polynomials $F_1, \ldots, F_m$ are *real*, *i.e.*, $F_i \in \mathbb{R}[x_0, \ldots, x_n]$ for $i = 1, \ldots, m$. For each point $x = [x_0 : \cdots : x_n] \in X$, its conjugate $\overline{x}$, whose coordinates are the complex conjugate of the coordinates of x, also occurs in X. Therefore X admits an antianalytic involution, namely

$$\sigma_X : X \to X : x \mapsto \overline{x},$$

which will be called *complex conjugation on X*. The pair (X, σ_X) is called a *smooth real algebraic curve*. Note that, as a consequence of the Implicit Function Theorem, a smooth real algebraic curve *is* a compact Klein surface. It is well known that the converse also holds.

Theorem 3. *Let (X, σ) be a compact Klein surface. Then there exists a smooth real algebraic curve (Y, σ_Y) such that*

$$(X, \sigma) \equiv (Y, \sigma_Y).$$

When (X, σ) is viewed as the smooth real algebraic curve (Y, σ_Y), its real part is precisely the set $Y(\mathbb{R})$ of points of Y with real coordinates:

$$Y(\mathbb{R}) = \{x = [x_0 : \cdots : x_n] \in \mathbb{P}^n(\mathbb{R}) : F_1(x) = \cdots = F_m(x) = 0\}.$$

Notice that $Y(\mathbb{R})$ may be empty. In particular, this shows that the classical correspondence between smooth complex algebraic curves and function fields in one variable over $\mathbb{C}$ does not work in a naive way in the real case. Indeed, from a family of real polynomials which define a smooth complex algebraic curve one also gets an algebraic function field in one variable over $\mathbb{R}$; however, its elements may not correspond to functions on the real zero set since it may be empty.

As we said before, there exist factors which obstructs two Klein surfaces from being isomorphic. The most obvious ones are of a topological nature. For example, if the Riemann surfaces X and Y have a different genus then the Klein surfaces (X, σ) and (Y, τ) cannot be isomorphic. Moreover, if X and Y have the same genus but the real parts X_σ and Y_τ have a different number of connected components, then (X, σ) and (Y, τ) are not homeomorphic and therefore cannot be isomorphic. (Two Klein surfaces are homeomorphic if there exists a homeomorphism between the Riemann surfaces compatible with the antianalytic involutions: see below.) This observation leads us to the following example.

Example 2. For $t = 0, 1$ consider the following homogeneous polynomial of degree 3

$$F_t(x, y, z) = y^2 z - x(x^2 + (-1)^t z^2).$$

An easy computation shows that at any point in $\mathbb{P}^2(\mathbb{C})$ the partial derivatives of F_t are not simultaneously zero, and so, each of the sets

$$X^t = \{[x : y : z] \in \mathbb{P}^2(\mathbb{C}) : F_t(x, y, z) = 0\}$$

is a compact Riemann surface. Since $F_t \in \mathbb{R}[x, y, z]$, we get two smooth real algebraic curves (X^t, σ_{X^t}) where σ_{X^t} denotes complex conjugation. The Riemann surfaces X^0 and X^1 are isomorphic, but the Klein surfaces (X^0, σ_{X^0}) and (X^1, σ_{X^1}) are not. Indeed, if ξ is a primitive 8th root of unity, then the map

$$\varphi : X^0 \to X^1 : [x : y : z] \mapsto [\xi x : \xi^4 y : \xi^3 z]$$

is an isomorphism of compact Riemann surfaces. However, $\varphi \circ \sigma_{X^0} \neq \sigma_{X^1} \circ \varphi$, and in fact, there is no isomorphism between X^0 and X^1 compatible with σ_{X^0} and σ_{X^1}. Indeed, the real part of (X^0, σ_{X^0}) has one connected component whereas that of (X^1, σ_{X^1}) has two, as can be easily seen by drawing their affine real parts.

$$\{(x,y) \in \mathbb{R}^2 : y^2 = x(x^2 + 1)\} \qquad\qquad \{(x,y) \in \mathbb{R}^2 : y^2 = x(x^2 - 1)\}$$

Thus (X^0, σ_{X^0}) and (X^1, σ_{X^1}) are not isomorphic.

The genus and the number of connected components of the real part are the first topological obstructions which prohibit two Klein surfaces from being isomorphic. However, they are not the only ones since both data do not determine the topology of a Klein surface. Indeed, it is well known that a third parameter is required.

Two Klein surfaces (X, σ) and (Y, τ) are said to be *topologically equivalent* if there exists a homeomorphism $\varphi : X \to Y$ such that $\varphi \circ \sigma = \tau \circ \varphi$. Weichold [23] proved in the last century that the topology of the Klein surface (X, σ) is determined by its *topological type*, which is the triple (g, k, ε) where

- g is the genus of X,

- k is the number of connected components of the real part X_σ and

- $\varepsilon = 0$ if $X - X_\sigma$ is not connected and $\varepsilon = 1$ otherwise.

For a proof of this result in modern terms see the article of Gromadzki [15] in this volume.

Obviously, isomorphic Klein surfaces are topologically equivalent, but the converse is not true, *i.e.*, there are non-isomorphic compact Klein surfaces with the same topological type. We shall give some examples. To do this, we must introduce the *automorphism group* of a Riemann surface X:

$$\text{Aut} X = \{\varphi : X \to X : \varphi \text{ is an analytic isomorphism}\}.$$

Its subgroup

$$\text{Aut}(X, \sigma) = \{\varphi \in \text{Aut} X : \varphi \circ \sigma = \sigma \circ \varphi\}$$

is called the *automorphism group of the Klein surface* (X, σ).

Observe that if X and Y (respectively (X, σ) and (Y, τ)) are isomorphic, then the groups $\mathrm{Aut}\,X$ and $\mathrm{Aut}\,Y$ (respectively $\mathrm{Aut}(X, \sigma)$ and $\mathrm{Aut}(Y, \tau)$) are isomorphic too. Thus, in some cases, we can use the automorphism group to decide that two Klein surfaces **are not** isomorphic. This leads us to the third example.

Example 3. Let X be the compact Riemann surface of genus 2 associated to the polynomial

$$F(x, y) = y^2 - (x^2 - 1)(x^2 - 4)(x^2 - 9) \in \mathbb{R}[x, y].$$

A picture of the set of zeroes of F in $\mathbb{R}^2$ is given in the next figure.

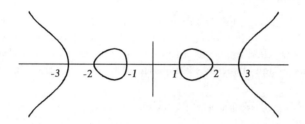

Hence, endowing X with the complex conjugation σ_X, we get the Klein surface (X, σ_X) which has topological type $(2, 3, 0)$ (the two unbounded components in the above affine picture join in the projective plane). On the other hand, for all values of $a, b \in \{+1, -1\}$, the isomorphism $\varphi_{a,b} : X \to X$ induced by the assignment

$$(x, y) \mapsto (ax, by)$$

commutes with σ_X. Therefore, the automorphism group $\mathrm{Aut}(X, \sigma_X)$ contains the subgroup of order 4 $\{\varphi_{a,b} : a, b = \pm 1\}$.

If we now choose six different real numbers $a_1, \ldots, a_6$ "at random", we get the compact hyperelliptic Riemann surface Y of genus 2 associated to the real polynomial

$$G(x, y) = y^2 - (x - a_1) \cdots (x - a_6).$$

Again, (Y, σ_Y) has topological type $(2, 3, 0)$. However, for most choices of $a_1, \ldots, a_6$, (Y, σ_Y) has no analytic automorphisms other than the identity and the *hyperelliptic involution*, namely, the automorphism $h_Y : Y \to Y$ induced by the assignment

$$(x, y) \mapsto (x, -y).$$

So, $\mathrm{Aut}(Y, \sigma_Y) = \{id, h_Y\}$ and the groups $\mathrm{Aut}(X, \sigma_X)$ and $\mathrm{Aut}(Y, \sigma_Y)$ are not isomorphic. Consequently, the compact Klein surfaces (X, σ_X) and (Y, σ_Y) are not isomorphic although they have the same topological type.

Since topological obstructions do not suffice to classify compact Klein surfaces under isomorphism, it is interesting to describe, for each triple (g, k, ε) corresponding to the topological type of some compact Klein surface, the *moduli space*

$$\mathcal{M}_{(g,k,\varepsilon)} = \left\{ \begin{array}{c} \text{isomorphism classes of compact Klein surfaces} \\ \text{of topological type } (g, k, \varepsilon) \end{array} \right\}.$$

These spaces are endowed with a quotient topology which make them connected. Furthermore, they admit a natural semianalytic structure [20], [16], [19]. In view of the above example, it is also interesting to describe, for each $\mathcal{M}_{(g,k,\varepsilon)}$, the subsets consisting of those surfaces which have the same automorphism group.

A detailed study of the moduli spaces $\mathcal{M}_{(g,k,\varepsilon)}$ appears in [11] for genus $g = 2$ and in [9] for $g = 3$. In the case of genus 2 there are 5 moduli spaces $\mathcal{M}_{(2,k,\varepsilon)}$. Each one is described in [11] as a semianalytic subset Δ of $\mathbb{R}^3$. Moreover, the subspaces consisting of surfaces having the same automorphism group are also determined.

As an example, we choose the case $k = \varepsilon = 1$.

Theorem 4. ([11],[12]) *Let Δ be the following subset of $\mathbb{R}^3$:*

$$\Delta = \{(a, b, c) \in \mathbb{R}^3 : b > 0, \ (a, b) \neq (c, 1), \ a \geq bc, \ a^2 + b^2 \geq c^2 + 1\}.$$

For each point $(a, b, c) \in \Delta$ let $X(a, b, c)$ be the hyperelliptic compact Riemann surface defined by the real polynomial

$$y^2 - x[(x - a)^2 + b^2][(x - c)^2 + 1].$$

Let σ_X denote complex conjugation on $X(a, b, c)$. The map

$$\Delta \to \mathcal{M}_{(2,1,1)} : (a, b, c) \mapsto [(X(a, b, c), \sigma_X)]$$

is a bijection, where $[(X(a, b, c), \sigma_X)]$ denotes the isomorphism class of the curve $(X(a, b, c), \sigma_X)$.

Moreover,

$$\mathrm{Aut}(X(a, b, c), \sigma_X) = \left\{ \begin{array}{ll} D_2 & \text{if} \ \ a = bc \ \ \text{or} \ \ a^2 + b^2 = c^2 + 1, \\ \mathbb{Z}_2 & \text{if} \ \ a > bc \ \ \text{and} \ \ a^2 + b^2 > c^2 + 1. \end{array} \right.$$

The next figure illustrates what Δ looks like. On it, the two shadowed surfaces correspond to curves with $\mathrm{Aut}(X(a, b, c), \sigma_X) = D_2$, while the open subset bounded by them correspond with curves with $\mathrm{Aut}(X(a, b, c), \sigma_X) = \mathbb{Z}_2$.

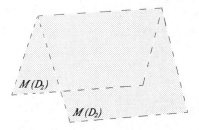

Even more, we can also compare the automorphism group $\mathrm{Aut}(X(a,b,c),\sigma_X)$ of the real curve $(X(a,b,c),\sigma_X)$ with the automorphism group $\mathrm{Aut}\,X(a,b,c)$ of the complex curve $X(a,b,c)$. In fact, denoting $p=(a,b,c)$ for short, we have:

(1) If $a=bc$ then $\mathrm{Aut}(X(p),\sigma_X)=D_2$ and

$$\mathrm{Aut}\,X(p) = \begin{cases} G_{24} & \text{if } c=0 \text{ and } b=3, \\ D_4 & \text{if } c=0 \text{ and } b\neq 3, \\ D_6 & \text{if } c^2>3 \text{ and } b=\frac{c^2+1}{c^2-3}, \\ D_2 & \text{otherwise,} \end{cases}$$

where $G_{24}=\langle f,g \mid f^4=g^6=(f\circ g)^2=(f^{-1}\circ g)^2=id\,\rangle$.

(2) If $a^2+b^2=c^2+1$ then $\mathrm{Aut}(X(p),\sigma_X)=D_2$ and

$$\mathrm{Aut}\,X(p) = \begin{cases} G_{48} & \text{if } b=1 \text{ and } c=-1, \\ D_4 & \text{if } b=1 \text{ and } c\neq -1, \\ D_2 & \text{otherwise,} \end{cases}$$

where $G_{48}=\langle f,g \mid f^8=g^3=(f\circ g)^2=(f^{-3}\circ g)^2=id\,\rangle$.

(3) If $a>bc$ and $a^2+b^2>c^2+1$ then $\mathrm{Aut}(X(p),\sigma_X)=Z_2$ and

$$\mathrm{Aut}\,X(p) = \begin{cases} D_6 & \text{if } c\in(-\sqrt{3},-1) \text{ and } \frac{c^2+1}{c^2-3}=-b=\frac{a}{c} \text{ or} \\ & \quad c\in(-1,0) \text{ and } \frac{c^2-3}{c^2+1}=-b=\frac{a}{c}, \\ Z_{10} & \text{if } (a,b,c)=(2\sin\frac{2\pi}{5},2\cos\frac{2\pi}{5},\tan\frac{\pi}{5}) \text{ or} \\ & \quad (a,b,c)=(-\sqrt{5+2\sqrt{5}},2+\sqrt{5},-\sqrt{5+2\sqrt{5}}), \\ D_2 & \text{if } a=-bc \text{ and } \frac{1+c^2}{3-c^2}\neq b\neq\frac{3-c^2}{1+c^2}, \\ Z_2 & \text{otherwise.} \end{cases}$$

In all cases, the explicit formulae for the generators of these groups are known [11], [12].

We are now in a position to provide two new examples.

Example 4. Let (X, σ_X) and (Y, σ_Y) be the hyperelliptic compact Klein surfaces corresponding, respectively, to the real polynomials

$$F(x,y) = y^2 - x[(x-10)^2 + 25][(x-2)^2 + 1]$$

and

$$G(x,y) = 80y^2 - x[20x^2 - 44x + 121][4x^2 + 4x + 5].$$

With the notations above, they are

$$(X, \sigma_X) = (X(10,5,2), \sigma_X) \quad \text{and} \quad (Y, \sigma_Y) = (X(\tfrac{11}{10}, \tfrac{11}{5}, \tfrac{-1}{2}), \sigma_X).$$

Both have topological type $(2,1,1)$ and the same complex automorphism group $\mathrm{Aut} X = \mathrm{Aut} Y = D_6$. However, the real automorphism groups $\mathrm{Aut}(X, \sigma_X)$ and $\mathrm{Aut}(Y, \sigma_Y)$ do not coincide. Therefore the Klein surfaces (X, σ_X) and (Y, σ_Y) are not isomorphic.

We can find non-isomorphic compact Klein surfaces with more common features.

Example 5. Let (X, σ_X) and (Y, σ_Y) be the hyperelliptic compact Klein surfaces corresponding, respectively, to the real polynomials

$$F(x,y) = y^2 - x(x^2+1)(x^2+4) \quad \text{and} \quad G(x,y) = y^2 - x[(x-2)^2+1][(x+2)^2+1].$$

With the notations in Theorem 4,

$$(X, \sigma_X) = (X(0,2,0), \sigma_X) \quad \text{and} \quad (Y, \sigma_Y) = (X(2,1,-2), \sigma_X).$$

Thus, both have topological type $(2,1,1)$, the same complex automorphism group $\mathrm{Aut} X = \mathrm{Aut} Y = D_4$, and also the same real automorphism group $\mathrm{Aut}(X, \sigma_X) = \mathrm{Aut}(Y, \sigma_Y) = D_2$. However, both compact Klein surfaces (X, σ_X) and (Y, σ_Y) are not isomorphic since they come from different points $p = (0,2,0)$ and $q = (2,1,-2)$ of Δ.

Remark. After some computations it can be proved that the Riemann surfaces X and Y in the last two examples are not isomorphic. This leads us naturally to look for non-isomorphic Klein surfaces (X, σ) and (Y, τ) such that X and Y are isomorphic Riemann surfaces. Equivalently, the problem can be formulated as follows: find, for a fixed compact Riemann surface X, all non-isomorphic Klein surfaces of the form (X, σ). Such Klein surfaces are called *real forms of X*. This problem was introduced in [18], and there exists a vast literature concerning it, mostly devoted to two problems:

i) to count the number of non-isomorphic Klein surfaces (X, σ);

ii) to compute the topological type of them.

We refer the interested reader to the papers [8], [6], [4], [14], [7] and [3] among others. Except in the last paper, which involves the classical theory of valuations and the subtleties of the real case, the arguments employed are combinatorial, using uniformization of Riemann and Klein surfaces by means of Fuchsian and non-euclidean crystallographic groups respectively. For a thorough treatment of this combinatorial point of view the reader is refered to the article [15] in this volume.

As we shall see, answers to both problems *i*) and *ii*) are very useful to solve a third one:

iii) to calculate "equations" for the Klein surfaces (X, σ).

What do we mean? In principle, to find equations for the Riemann surface X, to give the "formula" of each antianalytic involution σ and to determine the topological type of the Klein surface (X, σ).

There are few results in this direction, among them we mention here [21], [5] and [12]. In all cases, a fundamental tool to attack the problem is the knowledge of the automorphism group Aut$X$ of X. The reason is that its elements may be seen as automorphisms of the field of meromorphic functions of X. We explain this by developing another example.

Example 6. Let X be the hyperelliptic compact Riemann surface corresponding to the real polynomial

$$F(x, y) = y^2 - x^6 + 1.$$

Of course, (X, σ_X) (σ_X refers to complex conjugation) is a real form of X, and it is not difficult to see that its topological type is (2,1,1). Combinatorial methods allow us to know (see [5] or [15]) that X has exactly four (non-isomorphic) real forms, and that their topological types are

$$(2, 1, 1), \quad (2, 1, 0), \quad (2, 3, 0) \quad \text{and} \quad (2, 0, 1).$$

This has also been obtained in [12] in a geometric way.

We have already found the first real form and we look for the others. To this purpose, a fundamental tool is to know its automorphism group Aut$X$. It was computed independently by Accola in 1968 [1] and Maclachlan in 1969 [17]. It has presentation

$$\text{Aut}X = \langle f, g \mid f^6 = g^4 = (fg)^2 = fg^2 f^{-1} g^2 = 1 \rangle.$$

We think of these automorphisms as $\mathbb{C}$-automorphism of the field E of meromorphic functions of X. The reason is that non-isomorphic real forms of X

correspond to non-conjugate *symmetries* of E. In our case, E is the finitely generated extension $E = \mathbb{C}(x, y)$ of $\mathbb{C}$ by two trascendental (over $\mathbb{C}$) elements x and y related by

$$y^2 = x^6 - 1.$$

Let ξ be a primitive 6th root of unity. Then the $\mathbb{C}$-automorphisms α and β defined by

$$\begin{array}{ccc}
\alpha : & E & \to & E \\
& x & \mapsto & \xi x \\
& y & \mapsto & -y
\end{array} \quad \text{and} \quad \begin{array}{ccc}
\beta : & E & \to & E \\
& x & \mapsto & 1/x \\
& y & \mapsto & iy/x^3,
\end{array}$$

generate Aut$X$ since they satisfy the required relations

$$\alpha^6 = \beta^4 = (\alpha\beta)^2 = \alpha\beta^2\alpha^{-1}\beta^2 = 1.$$

We now look for $\mathbb{R}$-automorphisms of E which represent antianalytic involutions on X. Of course, the complex conjugation σ_X on X is represented by the automorphism σ which fixes the generators of E and whose restriction to $\mathbb{C}$ is the complex conjugation, that is:

$$\begin{array}{ccc}
\sigma : & E & \to & E \\
& i & \mapsto & -i \\
& x & \mapsto & x \\
& y & \mapsto & y
\end{array}$$

It is very easy to check that the $\mathbb{R}$-automorphisms

$$\tau_1 = \sigma, \quad \tau_2 = \alpha \circ \sigma, \quad \tau_3 = \beta \circ \sigma \quad \text{and} \quad \tau_4 = \alpha \circ \beta^2 \circ \sigma$$

have also order two. Since they are words in α, β and σ with and odd number of σ's, they represent antianalytic involutions on X. Their "formulae" are:

$$\begin{array}{ccc}
\tau_2 : & E & \to & E \\
& i & \mapsto & -i \\
& x & \mapsto & \xi x \\
& y & \mapsto & -y
\end{array} \qquad \begin{array}{ccc}
\tau_3 : & E & \to & E \\
& i & \mapsto & -i \\
& x & \mapsto & 1/x \\
& y & \mapsto & iy/x^3
\end{array} \qquad \begin{array}{ccc}
\tau_4 : & E & \to & E \\
& i & \mapsto & -i \\
& x & \mapsto & \xi x \\
& y & \mapsto & y.
\end{array}$$

We denote the associated antianalytic involution of X by the same letter τ_j. In this way we have four real forms (X, τ_j) and we examine whether (X, τ_i) and (X, τ_j) are isomorphic or not for $i \neq j$. This is equivalent to deciding if there exists a $\varphi \in$ Aut$X$ such that

$$\varphi \circ \tau_i = \tau_j \circ \varphi,$$

i.e., to deciding if τ_i and τ_j are conjugate with respect to the group Aut$X$. This is an accessible task for either group theoretists or even computers, but we prefer to argue in a different way. For each τ_i we shall find a smooth real algebraic curve (X_i, σ_i) (where σ_i is complex conjugation) isomorphic to (X, τ_i). This will help us to compute the topological type of (X, τ_i). For τ_1 we obviously take

$$\sigma_1 = \tau_1 \quad \text{and} \quad X_1 = X.$$

For the others, it suffices to find two elements $u(x, y)$ and $v(x, y)$ in E which are fixed by τ_i such that the field extension $E \mid \mathbb{R}(u, v)$ has degree two. In fact, if this is the case there exists a $\mathbb{C}$- isomorphism φ between $\mathbb{C}(u, v)$ and $\mathbb{C}(x, y)$ such the conjugate σ_i of τ_i by φ fixes the generators u and v of $\mathbb{C}(u, v)$ and whose restriction to $\mathbb{C}$ is complex conjugation. Therefore σ_i represents complex conjugation.

For $i = 2$, notice that $u = (1 + \xi)x$ and $v = iy$ are fixed by τ_2. Since

$$x = \frac{u}{1 + \xi} \quad \text{and} \quad y = -iv$$

it follows that $E = \mathbb{C}(u, v)$, and so $[E : \mathbb{R}(u, v)] = 2$. Moreover, if we substitute the values of x and y in the polynomial

$$F_1(x, y) = F(x, y) = y^2 - x^6 + 1$$

defining X, then one gets the polynomial

$$F_2(u, v) = v^2 - \lambda u^6 - 1,$$

which is real since $\lambda = -1/(1 + \xi)^6$ is a real number, positive, in fact. Now, we take as X_2 the hyperelliptic compact Riemann surface defined by F_2. Clearly, (X_2, σ_2) is isomorphic to (X, τ_2) via the isomorphism

$$\varphi : X \to X_2 : (x, y) \mapsto ((1 + \xi)x, iy).$$

It is not hard to see that (X_2, σ_2), and so (X, τ_2), has topological type $(2, 1, 0)$. Hence, the real forms (X, τ_1) and (X, τ_2) are not isomorphic.

Determining an equation which expresses τ_3 as complex conjugation is a little more difficult. It is straightforward to check that $u = (x + i)/(ix + 1)$ is fixed by τ_3. In order to find another element fixed by τ_3 and its relation with u we view u as a Möbius transformation on x

$$u = \psi(x) = \frac{x + i}{ix + 1},$$

and use the known relation between x and y. Note that ψ maps i, $-i$ and 1 onto ∞, 0 and 1 respectively, and so the unit circle of $\mathbb{C}$ onto the real line $\mathbb{R} \cup \{\infty\}$.

The expression of x in terms of u is

$$x = \phi(u) = \psi^{-1}(u) = \frac{u - i}{1 - iu}.$$

Substituting $x = \phi(u)$ in the expression $x^6 - 1$ we obtain

$$\phi(u)^6 - 1 = \frac{P(u)}{(1 - iu)^6},$$

where $P(u) = (u - i)^6 - (1 - iu)^6$ is a polynomial whose roots are all real. Indeed, if a is a root of P then $\phi(a) = \psi^{-1}(a)$ is a 6th root of unity. But ψ maps such roots onto the real line; so $a = \psi(\phi(a))$ belongs to $\mathbb{R} \cup \{\infty\}$, and in fact, $a \in \mathbb{R}$ since $\psi^{-1}(\infty) = i$ which is not a 6th root of unity. Hence, since the leading coefficient of P is 2, we get

$$P(u) = 2 \prod_{j=1}^{6} (u - a_j) \quad \text{with} \quad a_j \in \mathbb{R}.$$

On the other hand, in terms of x and y,

$$P(u) = (\phi(u)^6 - 1)(1 - iu)^6 = (x^6 - 1)(1 - i\psi(x))^6 = y^2(1 - i\psi(x))^6.$$

Therefore,

$$\prod_{j=1}^{6} (u - a_j) = \frac{y^2(1 - i\psi(x))^6}{2} = v^2,$$

where $v = y(1 - i\psi(x))^3/\sqrt{2} = \sqrt{32}y/(1 + ix)^3$ is an element fixed by τ_3. This is the element we are looking for. In view of its relation with u we take X_3 as the hyperelliptic compact Riemann surface defined by the real polynomial

$$F_3(u, v) = v^2 - \prod_{j=1}^{6} (u - a_j).$$

Clearly, (X_3, σ_3) is isomorphic to (X, τ_3) via the isomorphism

$$\varphi : X \to X_3 : (x, y) \mapsto (\psi(x), \sqrt{32}y/(1 + ix)^3).$$

The topological type of (X_3, σ_3), and so that of (X, τ_3), is (2,3,0). In fact, a picture of the set of real zeroes of F_3 is the following:

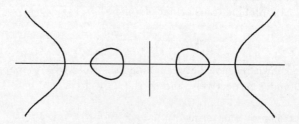

Hence, the real form (X, τ_3) is not isomorphic to neither (X, τ_1) nor (X, τ_2).

To finish, observe that τ_4 fixes $u = (1+\xi)x$ and $v = y$. Since $y^2 - x^6 + 1 = 0$, we get after substitution, $v^2 + \lambda u^6 + 1 = 0$, with λ as in the second case. In other words, the hyperelliptic compact Riemann surface X_4 defined by the real polynomial

$$F_4(u, v) = v^2 + \lambda u^6 + 1$$

together with the complex conjugation σ_4, is a Klein surface isomorphic to (X, τ_4) via

$$\varphi : X \to X_4 : (x, y) \mapsto ((1 + \xi)x, y).$$

Moreover, its real part is empty, because F_4 has no real solutions. So the topological type of (X, τ_4) is $(2,0,1)$.

This way we have found real equations for all of the real forms of the compact Riemann surface X.

A peculiarity of the last example is that the topological types of all (non-isomorphic) real forms of X are different, but this is not always the case. We finish with an example of such a situation [12].

Example 7. Let $a \neq 1$ be a positive real number, and let $X(a)$ be the compact Riemann surface associated to the real polynomial

$$F^a(x, y) = y^2 - x(x^2 + 1)(x - a)(x + a^{-1}).$$

It has exactly 4 non-isomorphic real forms. Two of them $(X_1(a), \sigma_1)$ and $(X_2(a), \sigma_2)$ have the same topological type $(2,2,1)$ and the same automorphism group D_4. $X_1(a)$ is defined by

$$F_1^a(x, y) = y^2 - x(x^2 + 1)(x - a)(x + a^{-1})$$

and $X_2(a)$ by

$$F_2^a(x, y) = y^2 - x(x^2 + 1)(x + a)(x - a^{-1}).$$

The other two $(X_3(a), \sigma_3)$ and $(X_4(a), \sigma_4)$ have the same topological type $(2,1,1)$ and the same automorphism group D_2. X_3 is defined by

$$F_3^a(x, y) = y^2 - x(x^2 - e^{i\alpha})(x^2 - e^{-i\alpha})$$

and $X_4(a)$ by

$$F_4^a(x, y) = y^2 - x(x^2 + e^{i\alpha})(x^2 + e^{-i\alpha}),$$

where $\cos \alpha = (a^2 - 1)/(a^2 + 1)$. In all cases σ_i is the complex conjugation on $X_i(a)$.

Note that $(X_1(a), \sigma_1)$ and $(X_2(a), \sigma_2)$ are real forms of the same Riemann surface with the same topological type and also with the same automorphism group. However they are not isomorphic. This also happens with $(X_3(a), \sigma_3)$ and $(X_4(a), \sigma_4)$.

We point out that this example presents another interesting fact: the surface $X(a)$ has even genus and four real forms with non-empty real part. This achieves the upper bound obtained recently in [14] for the number of real forms with non-empty real part that a compact Riemann surface of even genus may admit. A proof of this fact may also be found in this volume, see [15].

References

[1] R. D. M. ACCOLA, On the number of automorphisms of a closed Riemann surface, *Trans. Amer. Math. Soc.,* **131**, (1968), 398-408.

[2] N. L. ALLING, N. GREENLEAF, *Foundations of the Theory of Klein Surfaces.* Lecture Notes in Math., 219, Springer, 1971.

[3] M. E. ALONSO, P. VÉLEZ, On real involutions and ramification of real valuations, to appear in *Contemporary Mathematics.*

[4] S. A. BROUGHTON, E. BUJALANCE, A. F. COSTA, J. M. GAMBOA, G. GROMADZKI, Symmetries of Riemann surfaces on which PSL$(2, q)$ acts as Hurwitz automorphism group, *J. Pure Appl. Alg.,* **106**, (1996), 113-126.

[5] S. A. BROUGHTON, E. BUJALANCE, A. F. COSTA, J. M. GAMBOA, G. GROMADZKI, Symmetries of Accola-Maclachlan and Kulkarni surfaces, *Proc. Amer. Math. Soc.,* **127**, (1999), 637-646.

[6] E. BUJALANCE, A. F. COSTA, J. M. GAMBOA, Real parts of complex algebraic curves, Lecture Notes in Math. 1420, Springer-Verlag, Berlin, (1990), 749-759.

[7] E. BUJALANCE, G. GROMADZKI, M. IZQUIERDO, On real forms of a complex algebraic curve, preprint, 1998.

[8] E. BUJALANCE, D. SINGERMAN, The symmetry type of a Riemann Surface, *Proc. London Math. Soc.,* (3), **51**, (1985), 501-519.

[9] Y. CHECKOURI, *Périodes des courbes algébriques de genre* $g \leq 3$, Ph. D., Montpellier, 1997.

[10] C. CHEVALLEY, *Algebraic Function Fields in one Variable,* Math. Surveys **6**, Amer. Math. Soc., New York (1951).

[11] F. J. CIRRE, On the moduli of real algebraic curves of genus 2, preprint.

[12] F. J. CIRRE, Complex automorphism groups of real algebraic curves of genus 2, to appear in *J. Pure Appl. Alg.*

[13] J. M. GAMBOA, *Compact Klein Surfaces with boundary viewed as real compact smooth algebraic curves*, Mem. R. Acad. Ci. Madrid, **XXVII**, (1991).

[14] G. GROMADZKI, M, IZQUIERDO, Real forms of a Riemann surface of even genus, *Proc. Amer. Math. Soc.*, (12), **126**, (1998), 3475-3479.

[15] G. GROMADZKI, Symmetries of Riemann surfaces from a combinatorial point of view, this volume.

[16] J. HUISMAN, Real quotient singularities and nonsingular real algebraic curves in the boundary of the moduli space, *Compositio Mathematica*, **118**, (1), 43-60, 1999.

[17] C. MACLACHLAN, A bound for the number of automorphisms of a compact Riemann surface, *J. London Math. Soc.*, **44**, (1969), 265-272.

[18] S. M. NATANZON, On the order of a finite group of homeomorphisms of a surface into itself and the number of real forms of a complex algebraic curve, *Dokl. Akad. Nauk SSSR*, **242**, (1978), 765-768. (Soviet Math. Dokl., **19**, (5), (1978), 1195-1199.)

[19] M. SEPPÄLÄ, Moduli spaces of real algebraic curves, this volume.

[20] M. SEPPÄLÄ, R. SILHOL, Moduli spaces for real algebraic curves and real abelian varieties, *Math. Z.*, **201**, (1989), 151-165.

[21] P. TURBEK, The full automorphism group of the Kulkarni surface, *Rev. Mat. Univ. Complutense Madrid*, **10**, 2, (1997), 265-276.

[22] P. TURBEK, Compact Riemann surfaces and algebraic function fields, this volume.

[23] G. WEICHOLD, *Über symmetrische Riemannsche Flächen und die Periodizitätsmodulen der zugerhörigen Abelschen Normalintegrale erstes Gattung*, Dissertation, Leipzig, 1883.

Dept. de Geometría y Topología
Univ. Complutense de Madrid
jcirre@eucmos.sim.ucm.es

Dept. de Algebra
Univ. Complutense de Madrid
jmgamboa@eucmax.sim.ucm.es

MODULI SPACES OF REAL ALGEBRAIC CURVES

MIKA SEPPÄLÄ

1. INTRODUCTION

Let g, $g > 1$, be an integer, $k = \mathbb{R}$ or $k = \mathbb{C}$, and consider the moduli space

$$M_k^g = \{\text{isomorphism classes of genus } g \text{ algebraic curves over } k\}.$$

The complex moduli space $M_{\mathbb{C}}^g$ has been an object of intensive studies for more than 100 years now. It is known that $M_{\mathbb{C}}^g$ is a normal complex space and a quasiprojective algebraic variety, and the structure of $M_{\mathbb{C}}^g$ is, by now, fairly well understood in comparison with that of $M_{\mathbb{R}}^g$.

The moduli space of real algebraic genus g curves, $M_{\mathbb{R}}^g$, is the object of this review. Some properties of $M_{\mathbb{R}}^g$ can be derived from those of $M_{\mathbb{C}}^g$ but $M_{\mathbb{R}}^g$ is inherently more complex than $M_{\mathbb{C}}^g$, and new methods are needed. For instance, $M_{\mathbb{R}}^g$ does not have an algebraic structure while $M_{\mathbb{C}}^g$ does. Topologically $M_{\mathbb{R}}^g$ is more complicated than $M_{\mathbb{C}}^g$ — the real moduli space is not even connected.

The main result is that $M_{\mathbb{R}}^g$ has $\lfloor \frac{3g+4}{2} \rfloor$ components each of which is a real analytic subset of a complex orbifold. The moduli space $M_{\mathbb{R}}^g$ has a natural compactification obtained by allowing the curves to have double points in such a way that each component of the complement of the double points has a negative Euler characteristic. The moduli space $\overline{M}_{\mathbb{R}}^g$ of these stable genus g real algebraic curves is connected — a property conjectured by Klein.

The aim of this review is to explain the above results. To that end it is necessary to recall results about the topology and the geometry of real algebraic curves. That is done in Sections 2 and 3. These results about the topology of real algebraic curves are needed to count the number of components of $M_{\mathbb{R}}^g$, while the geometric results are needed to understand possible degenerations of real algebraic curves, i.e., to study the compactification $\overline{M}_{\mathbb{R}}^g$ of the real moduli space.

Components of $M_{\mathbb{R}}^g$ will be studied using quasiconformal mappings and classical theory of Teichmüller spaces. An effort has been made, in Sections 4 and 5, to explain these tools. Since these deliberations might not be detailed enough for a novice reader, references to all quoted results are included.

The inclusion $\mathbb{R} \subset \mathbb{C}$ induces a mapping $f : M_{\mathbb{R}}^g \to M_{\mathbb{C}}^g$ which simply forgets the real structure of a real curve. This mapping is not one–to–one, but it is of some interest anyway. The complex space $M_{\mathbb{C}}^g$ has a natural real structure which is induced by the complex conjugation. This real structure is an antiholomoprhic involution $\sigma^* : M_{\mathbb{C}}^g \to M_{\mathbb{C}}^g$ which takes the isomorphism class of a complex algebraic curve onto that of its complex conjugate. It is immediate that the image $f(M_{\mathbb{R}}^g) \subset M_{\mathbb{C}}^g$ is left point–wise fixed by this real structure, i.e., that $f(M_{\mathbb{R}}^g)$ is contained in the real part $M_{\mathbb{C}}^g(\mathbb{R})$ of $M_{\mathbb{C}}^g$. It was first observed by Earle that $f(M_{\mathbb{R}}^g) \neq M_{\mathbb{C}}^g(\mathbb{R})$, i.e., that there are complex algebraic curves which are isomorphic to their complex conjugates but are not isomorphic to curves defined by real polynomials. The presence of these complex curves makes the study of $f(M_{\mathbb{R}}^g)$ more complicated. It turns out that for $g > 3$, $f(M_{\mathbb{R}}^g)$ is that part of $M_{\mathbb{C}}^g(\mathbb{R})$ where the local real dimension of $M_{\mathbb{C}}^g(\mathbb{R})$ is as large as possible, i.e., $3g - 3$. Hence $f(M_{\mathbb{R}}^g)$ is the quasiregular real part of $M_{\mathbb{C}}^g$ and as such a semialgebraic subset of the quasiprojective complex variety $M_{\mathbb{C}}^g$.

This review does not contain new results. Most of the material presented here dates back to 1970's but some newer material is also included. Several of the results explained here were independently proved by mathematicians in the west and by mathematicians in the former Soviet Union where, in the 1970's, postdoctoral fellows and graduate students usually were not allowed to correspond with their colleagues in the West.

Main references to the theory reviewed here are [26] and [33]. A paper of Earle ([13]), in which he observed that $f(M_{\mathbb{R}}^g) \neq M_{\mathbb{C}}^g(\mathbb{R})$, was an important motivation for these studies. In that paper, Earle also developed Teichmüller space theory for real algebraic curves. That turned out to be an important tool for studying the components of $M_{\mathbb{R}}^g$.

Much of the material included in Sections 4 and 5 is explained in more detail in [22] and in [35].

By the real version of the Torelli theorem, one can study of $M_{\mathbb{R}}^g$ by studying the moduli space of principally polarized real abelian varieties, and vice versa. In that way it is possible to endow the components of $M_{\mathbb{R}}^g$ with semialgebraic structures, a result that appears to be out of reach of the analytic and geometric methods of this review. For details see [34] and Robert Silhol's paper ([37]) in this volume.

2. Topology of real algebraic curves

To fix the notation we recall that a topological model for a real algebraic curve C consists of an oriented compact genus g surface Σ together with an orientation reversing involution $\sigma : \Sigma \to \Sigma$. We assume here that $g > 1$. Real algebraic curves of the topological type (Σ, σ) are complex structures X on Σ such that $\sigma : (\Sigma, X) \to (\Sigma, X)$ is antiholomorphic. Note that whenever we consider a complex structure on an oriented surface Σ we assume that the orientation of this structure agrees with that of Σ.

For a real algebraic curve (Σ, X) we can also consider the associated *Klein surface* (cf. [2]) which is the quotient surface $X/\langle\sigma\rangle$. The fixed-points of the involution σ are boundary points of $X/\langle\sigma\rangle$. The Klein surface associated to a real algebraic curve may be non-orientable. In such a case it does not carry a complex structure but rather a so called dianalytic structure for which the coordinate transition functions are allowed to be either locally holomorphic or antiholomorphic.

Let
$$n = n(\Sigma, \sigma) = \text{ the number of the components of the fixed-point set } \Sigma_\sigma$$
and
$$k = k(\Sigma, \sigma) = 2 - \text{ the number of components of } \Sigma \setminus \Sigma_\sigma.$$
The topological invariants n and k together with the genus g of the surface Σ determine the topological type of (Σ, σ) uniquely. This means that if (Σ, τ) has the same invariants n and k as (Σ, σ), then there is a homeomorphism $f : \Sigma \to \Sigma$ such that $\sigma \circ f = f \circ \tau$. This also means that the quotient surfaces $\Sigma/\langle\sigma\rangle$ and $\Sigma/\langle\tau\rangle$ are homeomorphic. Here $\langle\sigma\rangle$ denotes the group generated by σ.

For a fixed genus g, the invariants n and k satisfy:

- $k = 0$ or $k = 1$
- If $k = 0$, then $0 < n \le g + 1$ and $n \equiv g + 1 \bmod 2$
- If $k = 1$, then $n \le g$.

These are the only restrictions, and there are $\lfloor \frac{3g+4}{2} \rfloor$ different topological types of real curves of genus g ([18], [19], [38]; for a recent detailed account see Paola Frediani's thesis [14]).

Let $(\Sigma_j, X_j, \sigma_j)$, $j = 1, 2$, be real algebraic curves. They are *real isomorphic* if there is a conformal mapping $h : (\Sigma_1, X_1) \to (\Sigma_2, X_2)$ such that $h \circ \sigma_1 = \sigma_2 \circ h$. This means that real isomorphic real algebraic curves are of the same topological type.

3. GEOMETRY OF REAL ALGEBRAIC CURVES

Let $X = (\Sigma, X)$ be a real algebraic curve of the topological type (Σ, σ). Assume that the genus of the compact topological surface Σ is at least 2. Recall that by the definition, $\sigma : X \to X$ is an antiholomorphic involution. Let $Z = X/\langle\sigma\rangle$ be the associated Klein surface.

Since the genus of the Riemann surface X is > 1, the unit disk D is the universal cover of the Riemann surface X. If G is the cover group, then $X = D/G$. Recall that G is a Fuchsian group consisting of (hyperbolic) Möbius transformations.

The involution $\sigma : X \to X$ lifts to an antiholomorphic self–mapping $s : D \to D$. The lifting s needs not be an involution anymore. It satisfies, however, the condition $s^2 \in G$. The mapping $s : D \to D$ is a glide reflection of D onto itself, i.e., a hyperbolic Möbius transformation followed by a reflection in its axis. R. J. Sibner ([36]) has analyzed in detail the different generating sets for the group $F = \langle G, s \rangle$ generated by the elements of G and by s.

We have

(1) $$X = D/G, \text{ and } Z = D/F.$$

Since the elements of F (and of G) are isometries of the unit disk equipped with the hyperbolic metric, the Riemann surface X and the Klein surface Z both get a hyperbolic metric from that of the unit disk. In this metric, the boundary components of the Klein surface Z are geodesic curves.

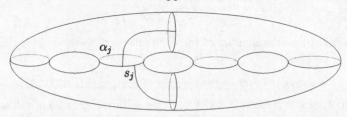

FIGURE 1.   The lengths of the decomposing curves α_j together
with the respective gluing angles θ_j are the Fenchel–Nielsen coordinates of a Riemann surface. The gluing angles are always
between 0 and 2π but the lengths of the decomposing curves
can be anything. The gluing angles are defined by dropping perpendicular geodesics to the curve α_j from adjacent decomposing
curves. An ordering as well as orientations of the curves are
needed to make this precise. Let s_j be the distance, measured in
the positive direction along the oriented curve α_j, between the
end–points of these perpendicular geodesics, and let ℓ_j denote
the length of α_j. Then the gluing angle θ_j is defined by $\theta_j = 2\pi s_j/\ell_j$.

Each compact Riemann or Klein surface of genus > 1 can be built by gluing
together *pairs of pants*, i.e., spheres with three geodesic boundary components.
Conversely, every such Riemann surface can be decomposed into pairs of pants
by disjoint simple closed geodesic curves α_j. The gluings of pairs of pants are
specified by so called gluing angles θ_j associated to each decomposing curve α_j.
These gluing angles θ_j together with the lengths of the decomposing geodesic
curves α_j form the *Fenchel–Nielsen parameters* of the Riemann surface X.
These parameters can be extended to Klein surfaces $Z = X/\langle\sigma\rangle$ with minor
technical modifications ([27]). For a general discussion with precise definitions
of the Fenchel–Nielsen coordinates see [39].

Lipman Bers has shown ([4]) that any hyperbolic compact Riemann surface
admits a length controlled decomposition into pairs of pants. In [33] Seppälä
extended that result and showed that, in the case of Riemann surfaces with
an antiholomorphic involution, i.e., in the case of real algebraic curves, this
decomposition can be so chosen that it is invariant under the involution (the
complex conjugation).

In [9] Buser and Seppälä derived the following linear bound:

Theorem 1. *Let X be a compact Riemann surface of genus g, $g > 1$. There
exist, on X, $3g - 3$ disjoint simple closed geodesic curves α_j of length $< 21g$.
These curves decompose X into pairs of pants. If X admits an antiholomorphic
involution $\sigma : X \to X$, then the curves α_j can be chosen in such a way that
the set of these curves is invariant under the mapping σ.*

The above bound, $21g$, for the lengths of the decomposing curves is much better than the one that the methods of Bers would yield. In a recent manuscript, Buser and Seppälä have extended these considerations to study "short" homology bases for Riemann and Klein surfaces ([10]).

This length controlled σ–invariant decomposition of a symmetric Riemann surface into pairs of pants has important applications in the study of compactifications of the moduli space of real algebraic curves.

4. QUASICONFORMAL MAPPINGS

Quasiconformal mappings provide convenient tools to study deformations of Riemann surfaces. In this Section we recall the their basic properties including the Teichmüller extremal mapping theorem.

We use the notations

$$\overline{\partial} = \frac{\partial}{\partial \overline{z}} = \frac{1}{2}\left(\frac{\partial}{\partial x} + i\frac{\partial}{\partial y}\right)$$

and

$$\partial = \frac{\partial}{\partial z} = \frac{1}{2}\left(\frac{\partial}{\partial x} - i\frac{\partial}{\partial y}\right).$$

Furthermore $\partial_\alpha f(z)$ denotes the derivative of f in the direction of a (unit) vector α.

Definition 1. *Let A be an open set in $\mathbb{C}$ and $K \geq 1$. A sense preserving homeomorphism $f : A \to f(A) \subset \hat{\mathbb{C}}$ is K-quasiconformal if and only if the following holds:*

- *f in absolutely continuous on lines in A.*
- $\max_\alpha |\partial_\alpha f(z)| \leq K \min_\alpha |\partial_\alpha(z)|$ *almost everywhere in A.*

A mapping f is quasiconformal *if it is K-quasiconformal for some constant K. The smallest number K for which a quasiconformal mapping f is K-quasiconformal, is called the* maximal dilatation *of f.*

Often quasiconformal mappings are defined using their more geometric properties (such as the fact that quasiconformal mappings map small circles onto small ellipses for which the ratio of the axes is bounded). Then the above definition is actually a theorem ([22, Theorem I.3.5]). For our purposes it is, however, convenient to regard this as the definition of quasiconformal mappings in the plane.

The property that f is absolutely continuous on lines guarantees that the mapping f has partial derivatives almost everywhere. Hence the second condition makes sense.

Let $f : A \to f(A)$ be a K-quasiconformal mapping and z a point where f is differentiable and $J_f(z) > 0$.

Since $\max_\alpha |\partial_\alpha f| = |\partial f| + |\overline{\partial} f|$, $\min_\alpha |\partial_\alpha f| = |\partial f| - |\overline{\partial} f|$, the second condition in the definition of quasiconformal mappings is equivalent to

(2) $$|\overline{\partial} f(z)| \leq \frac{K-1}{K+1}|\partial f(z)|.$$

Since the Jacobian

$$J_f(z) = |\partial f(z)|^2 - |\bar{\partial} f(z)|^2$$

is positive, $\partial f(z) \neq 0$, and we can form the quotient

$$(3) \qquad\qquad \mu_f(z) = \frac{\bar{\partial} f(z)}{\partial f(z)}.$$

Definition 2. *The function μ_f which is defined almost everywhere by the formula (3) is the complex dilatation of the quasiconformal mapping f.*

Observe:

- μ_f is a Borel measurable function.
- By (2), $|\mu_f(z)| \leq k = \frac{K-1}{K+1} < 1$ almost everywhere.

Definition 3. *A Borel measurable function $\mu : A \to \mathbb{C}$ which satisfies*

$$\operatorname{ess\,sup}_z \|\mu(z)\| < 1$$

is a Beltrami differential in the domain A. Let μ be a Beltrami differential in A. The differential equation

$$(4) \qquad\qquad\qquad \bar{\partial} f = \mu \partial f$$

is called a Beltrami equation.

For a holomorphic mapping f, μ_f vanishes identically, and (4) reduces to the Cauchy–Riemann equation $\bar{\partial} f = 0$.

Theorem 2. *A homeomorphism $f : A \to f(A)$ is K-quasiconformal if and only if the partial derivatives of f are locally in L^2 and satisfy the condition (2) for almost all z.*

Proof. [22, Theorem I.4.1].

Let f and g be quasiconformal mappings of a domain A with complex dilatations μ_f and μ_g, respectively. Direct computation yields the transformation rule

$$(5) \qquad \mu_{f \circ g^{-1}}(\zeta) = \frac{\mu_f(z) - \mu_g(z)}{1 - \mu_f(z)\overline{\mu_g(z)}} \left(\frac{\partial g(z)}{|\partial g(z)|} \right)^2, \quad \zeta = g(z),$$

which is valid for almost all $z \in A$, and hence for almost all $\zeta \in g(A)$.

The above transformation rule can be easily computed but it has deep consequences. Writing $(\partial g(z)/|\partial g(z)|)^2 = a$, $\mu_g(z) = b$, $\mu_f(z) = \xi$ and $\mu_{f \circ g^{-1}} = \omega$, the formula (5) becomes

$$\omega = \frac{a\xi - ab}{1 - \xi\bar{b}}.$$

The mapping $\xi \mapsto \omega$ is a Möbius–transformation mapping the unit disk onto itself. We conclude that *the complex dilatation of $\mu_{f \circ g^{-1}}$ depends holomorphically on the complex dilatation μ_f.* This fact plays a crucial role when defining the complex structure of a Teichmüller space.

The relationship between quasiconformal mappings and their Beltrami differentials or complex dilatations is very close. For our purposes they are simply two views of one object. That follows from the following important result.

Theorem 3 (Existence Theorem). *Let μ be a measurable function in a domain A. Assume that $\|\mu\|_\infty < 1$. There exists a quasiconformal mapping of A whose complex dilatation agrees with μ almost everywhere in A.*

Proof. [20, p. 136] or [21, p. 191].

The concept of quasiconformality extends naturally for Riemann and Klein surfaces.

Definition 4. *A homeomorphism $f : X \to X'$ between Klein surfaces is K-quasiconformal if it is locally K-quasiconformal in the following sense: for each point $P \in X$ there is a connected open neighborhood U of P, a dianalytic local variable $z : U \to z(U) \subset \mathbf{C}$ and a dianalytic local variable $w : V \to w(V)$ of a neighborhood V of $f(P)$ such that the mapping $w \circ f \circ z^{-1}$ is a K-quasiconformal homeomorphism of a neighborhood of $z(P)$.*

Computing locally we may form the Beltrami differential of a quasiconformal mapping $f : X \to X'$ between Klein surfaces: If $\{(U_i, z_i) \mid i \in I\}$ is a dianalytic atlas of X (whose charts U_i are connected), then for each index $i \in I$ there is a dianalytic chart (V_{j_i}, w_{j_i}) of X' such that $f(U_i) \subset V_{j_i}$ and $w_{j_i} \circ f \circ z_i^{-1}$ is quasiconformal in $z_i(U_i)$.

Let τ_i be the complex dilatation of this quasiconformal mapping. Consider the family of functions $\mu_i = \tau_i \circ z_i^{-1}$ associated to the different dianalytic charts of X. The function μ_i is a complex valued L^∞-function defined on U_i and $\|\mu_i\|_\infty \leq \frac{K-1}{K+1} < 1$.

To see how the different functions μ_i are related to each other, consider two intersecting dianalytic charts (U_i, z_i) and (U_j, z_j) of X. Define the function $T_{ij} : U_i \cap U_j \to \mathbf{C}$ setting

$$T_{ij} = (\partial(z_i \circ z_j^{-1}) + \overline{\partial}(z_i \circ z_j^{-1})) \circ z_j.$$

The following transformation formula is a straightforward computation: If $z_i \circ z_j^{-1}$ is holomorphic at $z_j(P)$, $P \in U_i \cap U_j$, then

$$(6) \qquad \mu_j(P) = \mu_i(P) \frac{\overline{T_{ij}(P)}}{T_{ij}(P)}.$$

If $z_i \circ z_j^{-1}$ is antiholomorphic at $z_j(P)$, then

$$(7) \qquad \overline{\mu_j(P)} = \mu_i(P) \frac{\overline{T_{ij}(P)}}{T_{ij}(P)}.$$

Definition 5. *We say that a collection $\mu = \{\mu_i \mid i \in I\}$ of measurable functions $\mu_i : U_i \to \mathbf{C}$ associated to dianalytic charts (U_i, z_i) of X is a $(-1,1)$-differential of the Klein surface X if the functions μ_i satisfy the transformation rules (6) and (7). If, in addition,*

$$\sup_{i \in I} \|\mu_i\|_\infty < 1$$

then μ is a Beltrami differential of the Klein surface X. *Let us use the notation* $D^{(-1,1)}(X)$ *for the space of (–1,1)–differentials of X and the notation $Bel(X)$ for the space of the Beltrami differentials of X.*

For a Riemann surface X, (i.e., for an orientable Klein surface X) $D^{(-1,1)}(X)$ is a complex Banach–space and $Bel(X)$ is its open unit ball. If X is not orientable, then, since the transformation rule (7) has to hold, elements of $Bel(X)$ cannot be multiplied by complex numbers. Hence, for a non–orientable Klein surface X, $Bel(X)$ is the open unit ball of a real Banach space.

By the above remarks, the complex dilatation of a quasiconformal mapping of a Klein surface X is a Beltrami differential of X. The following theorem is an immediate application of the existence and the uniqueness theorems of plane quasiconformal mappings:

Theorem 4. *Let μ be a Beltrami differential of a Klein surface $X = (\Sigma, X)$ then there exists a dianalytic structure X_μ of the topological surface Σ such that the identity mapping $(\Sigma, X) \rightarrow (\Sigma, X_\mu)$ is quasiconformal with the complex dilatation μ.*

If $f_i : X \rightarrow f_i(X)$, $i = 1, 2$, are two μ–quasiconformal mappings of X, then there exists an isomorphism $g : f_1(X) \rightarrow f_2(X)$ such that $f_2 = g \circ f_1$.

Proof. Let $\mathcal{U} = \{(U_i, z_i) \mid i \in I\}$ be a dianalytic atlas of (Σ, X). Let $\mu_i : U_i \rightarrow \mathbb{C}$ be the measurable function associated to the chart (U_i, z_i) by the Beltrami differential μ.

By Theorem 3 there exists $\mu \circ z_i^{-1}$–quasiconformal mappings $f_\mu^i : z_i(U_i) \rightarrow \mathbb{C}$. Choose one for each index $i \in I$.

Then by the transformation rule (5) $\mathcal{U}_\mu = \{(U_i, f_\mu^i \circ z_i \mid i \in I\}$ is dianalytic atlas of Σ. Let X_μ be the dianalytic structure defined by this atlas. By the definition it is now clear that the identity mapping $(\Sigma, X) \rightarrow (\Sigma, X_\mu)$ is μ–quasiconformal proving the first statement.

The second statement follows directly from the transformation rule (5). $\square$

Let $X = U/G$ where G is a reflection group. A (–1,1)–differential μ of X lifts to a function $\mu : U \rightarrow \mathbb{C}$. The transformation rules (6) and (7) are equivalent with the following formulae:

$$(8) \qquad \mu = (\mu \circ g)\frac{\overline{\partial g}}{\partial g}$$

for orientation preserving Möbius–transformations $g \in G$ and

$$(9) \qquad \overline{\mu} = (\mu \circ \sigma)\frac{\overline{\partial \sigma}}{\overline{\partial \sigma}}$$

for glide–reflections $\sigma \in G$.

Definition 6. *A measurable function μ satisfying the transformation rules (8) and (9) with respect to the elements of a reflection group G, is a (–1,1)–differential of the group G. If, in addition, $\|\mu\|_\infty < 1$, then μ is a*

Beltrami differential of the group G. *We use the notation* $D^{(-1,1)}(G)$ *for (– 1,1)-differentials of a reflection group* G, *and the notation* $Bel(G)$ *for Beltrami differentials of* G.

Provided that the group G does not contain orientation reversing elements $D^{(-1,1)}(G)$ is a complex Banach space and $Bel(G)$ is its open unit ball. If G contains also glide–reflections, $D^{(-1,1)}(G)$ is a real Banach space.

There is a real analytic homeomorphism between the unit disk and the complex plane. Nevertheless, there are no quasiconformal mappings between them. This follows rather easily from the quasi–invariance of conformal invariants under quasiconformal mappings. The case of compact Klein surfaces is, however, different: homeomorphic compact Klein surfaces are also quasiconformally equivalent by the following result.

Theorem 5. *Let X and Y be compact Klein surfaces. Each homeomorphism $f : X \to Y$ is homotopic to a quasiconformal mapping.*

Proof. It is well known that each homeomorphism $f : X \to Y$ is homotopic to a diffeomorphism. The dilatation quotient of such a diffeomorphism can be computed at each point of X. It does not depend on the choices of the local variables, and it is a continuous function on X. Since X is compact, this function has a finite maximum K on X. Hence a diffeomorphism homotopic to f is a quasiconformal mapping of X. There are many ways to find a diffeomorphism homotopic to a given homeomorphism. For more details and a direct construction we refer to [22, Theorem V.1.5.]. □

Let X and Y be homeomorphic compact Klein surfaces. By Theorem 5 there are quasiconformal mappings $X \to Y$. Teichmüller considered the problem of finding, in a given homotopy class of mappings $X \to Y$, one with the smallest maximal dilatation.

Theorem 6 (Teichmüller extremal mapping theorem). *Let X and Y be compact Riemann surfaces of genus $g > 1$ and $f : X \to Y$ a homeomorphism. There exists always a unique quasiconformal mapping $F : X \to Y$ such that the following holds:*

- *F is homotopic to the mapping f.*
- *If $g : X \to Y$ is a K_g-quasiconformal mapping and homotopic to f, then $K_g \geq K_F$, where K_g is the maximal dilatation of g and K_F is that of F. If $K_g = K_F$, then also $g = F$.*

Teichmüller gave also an explicit description of the geometry of such an extremal mapping F. The most delicate part of this result is the uniqueness of the extremal mapping. Detailed proofs for the above results of Teichmüller can be found, for instance, in [22, Chapter V].

Observe that non–classical Klein surfaces are not usually considered in this context but the above result holds also for them. In other words we have: *A homotopy class of a homeomorphism between compact non–classical compact Klein surfaces of genus $g > 1$ always contains a unique quasiconformal mapping having the smallest maximal dilatation.*

It is out of the scope of this presentation to prove these results. A clear exposition can be found, for instance, in the monograph of Olli Lehto [22]. Observe that, in the above theorems, the uniqueness part is not anymore true if we drop the assumption that the genus is at least two.

5. TEICHMÜLLER SPACES

5.1. Definitions. There are many alternative definitions for the Teichmüller space. We will next discuss some of these equivalent definitions.

Use the notation

$$\mathcal{M}(\Sigma) = \{\text{complex structures of } \Sigma \text{ which agree with the orientation of } \Sigma\}.$$

For a non–orientable surface $S = \Sigma/\langle\sigma\rangle$, we set

$$\mathcal{M}(S) = \{\text{dianalytic structures of } S\}.$$

Dianalytic structures differ from analytic structures in that the coordinate transition functions may also be antiholomorphic. Observe that the fixed–points of σ correspond to boundary points of S. The dianalytic structures of S are required to be such that they extend to the boundary. This implies that a dianalytic structure of S lifts to a usual complex structure of Σ.

The group $\mathrm{Homeo}(\Sigma)$ of homeomorphic self–mappings of Σ acts on $\mathcal{M}(\Sigma)$ in the following way. Let $\alpha \in \mathrm{Homeo}(\Sigma)$ and $X \in \mathcal{M}(\Sigma)$. Then $\alpha^*(X) \in \mathcal{M}(\Sigma)$ is defined requiring the mapping

$$(10) \qquad\qquad \alpha : (\Sigma, \alpha^*(X)) \to (\Sigma, X)$$

be either holomorphic or antiholomorphic depending on whether α is orientation preserving or not. In the same way we define the action of $\mathrm{Homeo}(S)$ on $\mathcal{M}(S)$.

It is clear, by the definitions, that the fixed-point set $\mathcal{M}(\Sigma)_{\sigma^*}$ consists of real algebraic curves of the topological type (Σ, σ).

Consider

$$\mathrm{Homeo}_0(\Sigma) = \{f \in \mathrm{Homeo}(\Sigma)| \, f \text{ homotopic to the identity}\},$$

and

$$\mathrm{Homeo}_0(S) = \{f \in \mathrm{Homeo}(S)| \, f \text{ homotopic to the identity}\},$$

Definition 7 (of Teichmüller spaces). *The set theoretic quotient*

$$T(\Sigma) = \mathcal{M}(\Sigma)/\mathrm{Homeo}_0(\Sigma)$$

is the Teichmüller space *of the surface Σ. We use also the notation T^g for the Teichmüller space of a genus g surface Σ. In the same way we define the Teichmüller space $T(S)$ of the surface $S = \Sigma/\langle\sigma\rangle$ as the quotient*

$$T(S) = \mathcal{M}(S)/\mathrm{Homeo}_0(S).$$

Observe that homotopic self–mappings of an oriented surface are simultaneously orientation preserving. Consequently, if Σ is oriented, then $\mathrm{Homeo}_0(\Sigma)$ is a subgroup of the group $\mathrm{Homeo}_+(\Sigma)$ of orientation preserving homeomorphic self–mappings of Σ.

Definition 8 (of the modular group). *For an oriented surface* Σ, *the group* $\Gamma(\Sigma) = \text{Homeo}_+(\Sigma)/\text{Homeo}_0(\Sigma)$ *is the* modular group *or the* mapping class group *of the surface* Σ. *For non–orientable surfaces* S the modular group *is defined setting* $\Gamma(S) = \text{Homeo}(S)/\text{Homeo}_0(S)$. *We use also the notation* Γ^g *for the modular group of a genus* g *surface* Σ.

Definition 9 (of the moduli space). *For an oriented surface* Σ, *the quotient* $M(\Sigma) = \mathcal{M}(\Sigma)/\text{Homeo}_+(\Sigma)$ *is the* moduli space . The moduli space *of a non–orientable surface* S *is* $M(S) = \mathcal{M}(S)/\text{Homeo}(S)$. *We use also the notation* M^g *for the moduli space of a smooth genus* g *surface* Σ.

It follows from the above definitions that the modular group acts on the Teichmüller space, and that $M(\Sigma) = T(\Sigma)/\Gamma(\Sigma)$. An orientation reversing self–mapping f of Σ induces, likewise, a mapping $f^* : T(\Sigma) \to T(\Sigma), [X] \mapsto [f^*(X)]$, where the analytic structure $f^*(X)$ of Σ is defined requiring (10) be antianalytic.

Let $X, Y \in \mathcal{M}(\Sigma)$ be two analytic structures of a fixed compact surface Σ. By Theorem 5 there are quasiconformal mappings $(\Sigma, X) \to (\Sigma, Y)$ homotopic to the identity mapping of Σ. For such a quasiconformal mapping f, let K_f denote the maximal dilatation of f.

Definition 10 (of the Teichmüller metric). *The distance between two points* $[X]$ *and* $[Y]$ *of the Teichmüller space* $T(\Sigma)$ *in the* Teichmüller metric τ *of the* $T(\Sigma)$ *is defined by*

$$\tau([X], [Y]) = \inf\{\frac{1}{2}\log K_f |\, f \in \text{Homeo}_0(\Sigma)\}.$$

The Teichmüller space $T(\Sigma)$ together with the Teichmüller metric is homeomorphic to an Euclidean space ([22, Theorem 9.2, page 241]). This is one of the deep consequences of the Teichmüller extremal mapping theorem that cannot be presented here. We refer to Chapter V of the monograph of Olli Lehto ([22]) for a complete and detailed treatment.

We observe, nevertheless, that the elements of the modular group $\Gamma(\Sigma)$ are isometries of the Teichmüller metric. This is an immediate consequence of the definitions. Hence they are, in particular, homeomorphic self–mappings of the Teichmüller space.

5.2. Teichmüller spaces of Beltrami differentials.

By Theorems 4 and 5 we can associate, to each Beltrami differential μ of a Klein surface X, a Klein surface X_μ, and vice versa. Consequently, Teichmüller spaces can be defined also in terms of Beltrami differentials.

To be more precise, choose a point $[X] \in T(\Sigma)$, which we will refer to as *the origin* of the Teichmüller space. Consider the space $Bel(X)$ of Beltrami differentials of X. Each $\mu \in Bel(X)$ defines a unique $X_\mu \in \mathcal{M}(\Sigma)$ such that the identity mapping of Σ is a $\mu-$ quasiconformal mapping $X \to X_\mu$. We say that two Beltrami differentials μ_1 and μ_2 of X are *equivalent*, $\mu_1 \approx \mu_2$, if the homotopy class of the identity mapping of Σ contains an isomorphism $(\Sigma, X_{\mu_1}) \to (\Sigma, X_{\mu_2})$.

Definition 11. *The set $T(X) = Bel(X)/\approx$ is* the Teichmüller space of Beltrami differentials of X.

It is an immediate consequence of these definitions that

(11) $$T(X) \to T(\Sigma), [\mu] \mapsto [X_\mu],$$

is a bijection between these two Teichmüller spaces.

The Teichmüller metric of $T(X)$ is the pull back of the Teichmüller metric of $T(\Sigma)$ under the mapping (11).

Theorem 7. *Teichmüller space $T(X)$ is connected.*

Proof. Let μ represent an arbitrary point of $T(X)$. Then $t \mapsto [t\mu]$ is a path connecting the point $[\mu]$ to the origin of $T(X)$. Note that this path depends on the choice of μ ([40]). $\square$

Lars Ahlfors showed as early as in 1959([1]) that, for an orientable surface Σ without boundary, the Teichmüller space $T(\Sigma)$ has a natural complex structure. The construction of Ahlfors was based on considering periods of Abelian differentials. It is not possible to present it here. We will, however, describe a way to decide which functions are holomorphic on $T(\Sigma)$. *Let us assume now that Σ is a compact and oriented surface without boundary.*

It is convenient to consider the Teichmüller space $T(X)$ of Beltrami differentials of X instead of $T(\Sigma)$. Let $\pi : Bel(X) \to T(X)$ be the projection. This is a continuous mapping with respect to the L^∞-metric on $Bel(X)$ and the Teichmüller metric on $T(X)$ (cf. e.g. [22, III.2.2]). Hence, for any open $U \subset T(X)$, $\pi^{-1}(U)$ is open in $Bel(X)$.

Holomorphic functions on $T(X)$. Let $U \subset T(X)$ be open. We declare a function $f : U \to \mathbb{C}$ *holomorphic* if the composition $f \circ \pi$ is a holomorphic function on the open set $\pi^{-1}(U)$ of the complex Banach space of $(-1,1)$–differentials of X.

In this way $T(X)$ becomes first *a ringed space*. It is not, *a priori*, clear that the above definition actually gives a good complex structure on $T(X)$. That is, however, the case if Σ is an oriented surface which does not have boundary (cf. e.g. [22, Chapter V]). This complex structure of $T(X)$ is then transported to a complex structure of $T(\Sigma)$ by requiring that the mapping (11) is holomorphic.

We still have to check that this complex structure of $T(\Sigma)$ does not depend on the choice of the origin X of the Teichmüller space $T(\Sigma)$. But that is an immediate consequence of the transformation formula (5) and the remark made after it.

The elements of the modular group are biholomorphic automorphisms of the Teichmüller space. *Royden has shown (*[29, Theorem 1 on page 281] *and* [30, Theorem 2 on page 379]*), in fact, that for surfaces of genus $g > 2$, $\Gamma(\Sigma)$ is the full group of holomorphic automorphisms of $T(\Sigma)$.* In the same way we verify that the mapping $\sigma^* : T(\Sigma) \to T(\Sigma)$, induced by an orientation reversing mapping $\sigma : \Sigma \to \Sigma$ (cf. formula (10)), is an antiholomorphic self–mapping of the Teichmüller space (for details see [31, 5.10]).

Theorem 8. *The Teichmüller space of compact genus g Riemann surfaces, $g > 1$, is a complex manifold of complex dimension $3g - 3$. The elements of the modular group Γ^g are holomorphic automorphisms of T^g. For $g > 2$, Γ^g is the full group of holomorphic automorphisms of T^g.*

This result can be extended to general finite dimensional Teichmüller spaces of oriented surfaces without boundaries. For a proof of this result we refer to the monograph of F. W. Gardiner [15, 9.2].

5.3. **Teichmüller spaces of real algebraic curves.** In order to study Teichmüller spaces of real algebraic curves of a given topological type (Σ, σ) we have two choices. We may consider the quotient surface $S = \Sigma/\langle\sigma\rangle$ and reproduce the classical theory there. Alternatively we may work on the oriented compact surface Σ and make the constructions σ-equivariantly. Both approaches lead to the same result.

Let us first consider $S = \Sigma/\langle\sigma\rangle$. In this case $T(S)$ is *not* going to be a complex manifold. The Teichmüller space is, instead, a real analytic manifold.

The surface Σ is a compact oriented surface without boundary together with a projection $\pi : \Sigma \to S$ that is a ramified double covering mapping. It is ramified precisely at the points lying over the boundary points of S. The covering group of π is generated by an orientation reversing involution σ of Σ for which $\pi \circ \sigma = \pi$. The fixed point set of the involution σ corresponds to the boundary points of S.

Above we have observed that the mapping $\sigma : \Sigma \to \Sigma$ induces an antiholomorphic self–mapping σ^* of $T(\Sigma)$. This mapping is an involution since σ is an involution.

For any $X \in \mathcal{M}(S)$ let $\pi^*(X)$ be the complex structure of Σ which agrees with the orientation of Σ and for which the projection

$$\pi : (\Sigma, \pi^*(X)) \to (S, X)$$

is dianalytic. It is immediate that

$$\pi^* : T(S) \to T(\Sigma), [X] \mapsto [\pi^*(X)],$$

is a well–defined mapping of Teichmüller spaces. It is not difficult to show that the mapping $\pi^* : T(S) \to T(\Sigma)$ is an isometry with respect to the corresponding Teichmüller metrics. Hence it is, in particular, a homeomorphism of $T(S)$ onto $\pi^*(T(S))$.

Theorem 9. *Assume that $S = \Sigma/\langle\sigma\rangle$ is a non–classical surface of genus g, $g > 1$. We have*

$$\pi^*(T(S)) = T(\Sigma)_{\sigma^*}$$

Proof. Let X be a complex structure of S. The complex structure $\pi^*(X)$ of Σ has the (defining) property that $\sigma : (\Sigma, \pi^*(X)) \to (\Sigma, \pi^*(X))$ is an antiholomorphic involution. This implies that

$$[\pi^*(X)] \in T(\Sigma)_{\sigma^*}, \text{ i.e. that } \pi^*(T(S) \subset T(\Sigma)_{\sigma^*}.$$

To prove the converse inclusion, take a point $[Y] \in T(\Sigma)_{\sigma^*}$. By the definition this means that there is a holomorphic mapping $f : (\Sigma, Y) \to (\Sigma, \sigma^*(Y))$ which is homotopic to the identity mapping of Σ. (Recall that $\sigma^*(Y)$ is defined as that complex structure of Σ for which the mapping $\sigma : (\Sigma, \sigma^*(Y)) \to (\Sigma, Y)$ is antiholomorphic.)

The construction implies that $\tau = \sigma \circ f : (\Sigma, Y) \to (\Sigma, Y)$ is an antiholomorphic mapping. Then $\tau^2 : (\Sigma, Y) \to (\Sigma, Y)$ is a holomorphic mapping. Since σ is an involution and f is homotopic to the identity, we conclude that τ^2 is also homotopic to the identity. Since the genus of the Riemann surface (Σ, Y) is at least 2, the only holomorphic automorphism of (Σ, Y) that is homotopic to the identity is the identity itself. This implies that $\tau : (\Sigma, Y) \to (\Sigma, Y)$ is an antiholomorphic *involution* proving the theorem. $\qquad\square$

For a surface $S = \Sigma/\langle \sigma \rangle$ that is either non–orientable or has a non–empty boundary (or both) we now identify $T(S)$ with its image $\pi^*(T(S)) = T(\Sigma)_{\sigma^*}$ in $T(\Sigma)$. Since σ^* is an antiholomorphic involution of $T(\Sigma)$, its fixed–point set, $\pi^*(T(\Sigma))$, is a real analytic manifold.

The Teichmüller spaces $T(S)$ were studied in the 1970's by several authors (C. Earle [13], S. Natanzon [26], and M. Seppälä [31]).

Let (Σ, σ) and (Σ, τ) be two different topological models for real algebraic curves of genus g. Then the respective Teichmüller spaces, $T(\Sigma)_{\sigma^*}$ and $T(\Sigma)_{\tau^*}$ are both real analytic manifolds. Since the topological types are different, the mappings σ^* and τ^*, of the Teichmüller space $T(\Sigma)$ onto itself, are not conjugate in the modular group. Buser and Seppälä (1998, [11]) have, however, constructed, for any such σ and τ, a real analytic diffeomorphism $d : T(\Sigma) \to T(\Sigma)$ such that $\sigma^* \circ d = d \circ \tau^*$. Thus any two real structures σ^* and τ^* of the Teichmüller space are conjugate in the group of real analytic diffeomorphisms of $T(\Sigma)$. This means, in particular, that the Teichmüller spaces of the respective real algebraic curves are diffeomorphic as real analytic manifolds (this can actually be shown also without this conjugating map d).

6. MODULI SPACES

6.1. Complex orbifolds.
To describe the structure of the various moduli spaces we next introduce the concept of *complex orbifolds* (cf. Ratcliffe [28, §13.2]).

Let G be the group of biholomorphic self–mappings of $\mathbb{C}^n$, and let M be a Hausdorff space. An $(\mathbb{C}^n, G)$–*orbifold chart* for M is a pair (U, ϕ) where

- $U \subset M$ is an open set, and
- ϕ is a homeomorphism of U onto $\phi(U) \subset \mathbb{C}^n/\Gamma_U$ where Γ_U a finite subgroup of G.

A collection of $(\mathbb{C}^n, G)$–orbifold charts

$$\mathcal{A} = \{(U_i, \phi_i : U \to \mathbb{C}^n/\Gamma_i) \mid i \in I\}$$

is an $(\mathbb{C}^n, G)$–*orbifold atlas* if

- $M = \cup_{i \in I} U_i$

- If U_i and U_j intersect, then the *coordinate change function*

$$\phi_j \circ \phi_i^{-1} : \phi_i(U_i \cap U_j) \to \phi_j(U_i \cap U_j)$$

has the property that if x and y are points of $\mathbb{C}^n$ such that

$$\phi_j(\phi_i^{-1}(\Gamma_i x)) = \Gamma_j y,$$

then there is an element g of G such that $gx = y$ and g lifts $\phi_j \circ \phi_j^{-1}$ in a neighborhood of x, i.e.,

$$\phi_j \circ \phi_i^{-1}(\Gamma_i w) = \Gamma_j g w$$

for all w in a neighborhood of x.

Locally, a complex orbifold M is the quotient of an open set in $\mathbb{C}^n$ mod the action of finite group of holomorphic automorphisms of $\mathbb{C}^n$. A function $f : M \to \mathbb{C}$ is a *holomorphic function of the complex orbifold* if, near every point, it lifts to a holomorphic mapping of an open set of $\mathbb{C}^n$ into $\mathbb{C}$. In the same way we define *real analytic* functions of a complex orbifold. The condition regarding the change of variable implies that the above definitions do not depend on the choice of the local variable.

6.2. Moduli spaces of real curves of a given topological type. The *moduli space* of the surface Σ is

(12) $$M(\Sigma) = \mathcal{M}(\Sigma)/\text{Homeo}_+(\Sigma),$$

where $\text{Homeo}_+(\Sigma)$ denotes the subgroup of the group $\text{Homeo}(\Sigma)$ consisting of those homeomorphisms which preserve the orientation. The *modular group* $\Gamma(\Sigma) = \text{Homeo}_+(\Sigma)/\text{Homeo}_0(\Sigma)$ acts on the Teichmüller space $T(\Sigma)$, and excluding certain special cases, $\Gamma(\Sigma)$ acts properly discontinuously as the full group of holomorphic automorphisms of $T(\Sigma)$ ([30]).

Consider the Teichmüller space $T(\Sigma)$. Let X be a complex stucture on Σ. For any choice of X, we can form the Bers' embedding ([3]) of $T(\Sigma)$ into the space of quadratic differentials on (Σ, X). The construction of this embedding is rather technical. What matters for us is that the space of quadratic differentials of a compact genus g Riemann surface is $\mathbb{C}^{3g-3}$. Bers' construction implies that the subgroup

$$\Gamma_X = \{\gamma \in \Gamma(\Sigma) \mid \gamma([X]) = [X]\}$$

of the modular group fixing the point $[X] \in T(\Sigma)$ corresponds to a group of linear mappings of the space of quadratic differentials of (Σ, X) onto itself. Also, since the group $\Gamma(\Sigma)$ acts properly discontinuously on $T(\Sigma)$, every group Γ_X is finite.

It then follows, from the definitions, that the moduli space $M(\Sigma) = T(\Sigma)/\Gamma(\Sigma)$ is a complex orbifold.

In the definition of the moduli space of real algebraic curves of a given topological type (Σ, σ) it is, as in the case of Teichmüller spaces, convenient to consider the quotient surface $S = \Sigma/\langle \sigma \rangle$ rather than working on Σ. The moduli space of the possibly non–orientable surface S is simply

(13) $$M(S) = \mathcal{M}(S)/\text{Homeo}(S).$$

The space $M(S)$ is also referred to as *the moduli space of Klein surfaces of the topological type S*.

Complications start at this point. As in the case of Teichmüller spaces, consider the mapping

(14) $\rho : M(S) \to M(\Sigma), [X] \mapsto [X^d]$

defined by the liftings of dianalytic structures X of $S = \Sigma/\langle \sigma \rangle$ to analytic structures X^d of Σ. This mapping just forgets the real structure of the real algebraic curve (i.e., point in $M(S)$) in question.

Some symmetric Riemann surfaces X^d admit several antiholomorphic involutions ([5], [6], [7], [8], [17], [23], [24], [25]), hence the mapping $\rho : M(S) \to M(\Sigma)$ is not one-to-one.

More precisely, the mapping ρ is not one–to–one for the following reason. Define first the modular group $\Gamma(S)$ of the possibly non–orientable surface S in the usual way by setting

$$\Gamma(S) = \mathrm{Homeo}(S)/\mathrm{Homeo}_0(S).$$

Then $\Gamma(S)$ acts on the Teichmüller space $T(S)$, and clearly $M(S) = T(S)/\Gamma(S)$. Any homeomorphic self–mapping f of the surface $S = \Sigma/\langle \sigma \rangle$ lifts to a unique orientation preserving homeomorphism F of Σ. This lifting satisfies $F \circ \sigma = \sigma \circ F$. In this way, elements of $\Gamma(S)$ lift to elements of $\Gamma(\Sigma)$, and we can map $\Gamma(S)$ into $\Gamma(\Sigma)$. It is immediate, that $\Gamma(S)$ becomes, in this way, the centralizer of σ^* in $\Gamma(\Sigma)$. In particular, the lifting of $\Gamma(S)$ is *not* the full modular group $\Gamma(\Sigma)$. Consequently, the mapping $\rho : M(S) \to M(\Sigma)$ is not going to be one–to–one.

Let $\Gamma(\Sigma, \sigma) = \{\gamma \in \Gamma(\Sigma) \mid \sigma^*\gamma = \gamma\sigma^*\}$ be the centralizer of σ^* in $\Gamma(\sigma)$. Then we may consider the intermediate moduli space $M(\Sigma, \sigma) = T(\Sigma)/\Gamma(\Sigma, \sigma)$. The diagram

$$
\begin{array}{ccc}
T(S) & \xrightarrow{\ \rho\ } & T(\Sigma) \\
\downarrow{\scriptstyle \pi} & & \downarrow{\scriptstyle \pi} \\
T(S)/\Gamma(S) = M(S) & \xrightarrow{\ \rho\ } & M(\Sigma, \sigma) = T(\Sigma)/\Gamma(\Sigma, \sigma)
\end{array}
$$

(15)

commutes, and now the induced mapping $\rho : M(S) \to M(\Sigma, \sigma)$ is one–to–one.

On the other hand, $M(\Sigma, \sigma) = T(\Sigma)/\Gamma(\Sigma, \sigma)$ is a quotient of a complex manifold via the action of a properly discontinuous subgroup $\Gamma(\Sigma, \sigma)$ of the modular group. In view of the considerations regarding the orbifold structure of the ordinary moduli space, it follows that $M(\Sigma, \sigma)$ is also a complex orbifold. Now the image of $M(S)$ in $M(\Sigma, \sigma)$ is a projection of a real analytic submanifold $T(\Sigma)_{\sigma^*}$ of $T(\Sigma)$. Hence it is a real analytic subset of the orbifold $M(\Sigma, \sigma)$ ([32]) in the sense that it is locally the zero set of a real analytic function on $M(\Sigma, \sigma)$.

The situation here is more complicated than what one realizes at first. These complications are illustrated by the following example (cf. [16]).

Consider the finite complex plane $\mathbb{C}$, and the real analytic subset $\mathbb{R}$ consisting of the real points. The group generated by the involution $s(z) = -z$ acts on $\mathbb{C}$. The quotient space $X = \mathbb{C}/\langle s \rangle$ is then an orbifold with a singularity at the origin. The projection is given by

$$(16) \qquad\qquad \mathbb{C} \to X, \; z \mapsto w = z^2.$$

A holomorphic or a real analytic function f in $\mathbb{C}$ defines a holomorphic or a real analytic function on $X = \mathbb{C}/\langle s \rangle$ if and only if $f \circ s = f$.

The real analytic subset $\mathbb{R}$ of $\mathbb{C}$ is the zero–set of the s invariant real analytic function $(\operatorname{Im} z)^2$. Here $\operatorname{Im} z$ denotes the imaginary part of the complex number z. The projection of $\mathbb{R} \subset \mathbb{C}$ in $X = \mathbb{C}/\langle s \rangle$ is the non–negative real line. It is the zero–set of the real analytic function $(\operatorname{Im} \sqrt{w})^2$. Observe that this is a real analytic function in X, it is not a real analytic function in $\mathbb{C}$. The quotient space X has a cone singularity at the origin. The projection of the real analytic subset $\mathbb{R} \subset \mathbb{C}$ has also a singularity at the origin. This time the singularity is a boundary point. The projection of $\mathbb{R}$ in X is, nevertheless, a real analytic subspace of the complex orbifold X.

Observe that, in this example, the mapping $z \mapsto z^2$ of the Riemann surface $\mathbb{C}$ onto $\mathbb{C}/\langle s \rangle$ is holomorphic but not biholomorphic. The mapping is not conformal at the origin. Hence, as Riemann surfaces, $\mathbb{C} \neq \mathbb{C}/\langle s \rangle$.

The same phenomena is present in the moduli spaces of real curves. They are real analytic subsets of certain complex orbifolds, but they do have, in general, boundary points or, rather, cone points. These singular points correspond to singular points of the complex orbifold (see [16]).

6.3. Moduli spaces of real curves of a given genus. Let $g > 1$ be an integer. Consider *the moduli space of real algebraic curves of genus g*

$$M_{\mathbb{R}}^g = \{\text{isomorphism classes of real algebraic curves of genus } g\}.$$

This space is our main object of interest.

By considerations of Section 2, there are $m(g) = \lfloor \frac{3g+4}{2} \rfloor$ different topological types of real curves of genus g.

By considerations of the preceding Section, the moduli space $M(S)$ of real curves of the topological type $S = \Sigma/\langle \sigma \rangle$ is a connected real analytic subset of a complex orbifold. It is clear that real curves of the same genus but of different topological types cannot be real isomorphic. It follows that $M_{\mathbb{R}}^g$ is the disjoint union of the $m(g)$ moduli spaces $M(S)$.

It is of some interest to consider the mapping

$$f : M_{\mathbb{R}}^g \to M^g$$

which simply forgets the real structure of a real curve, and maps the real isomorphism class of a real algebraic curve C onto the complex isomorphism class of the same curve (viewed as a complex curves).

We have

Theorem 10. *Smooth real algebraic genus g curves in the moduli space of complex genus g curves form a connected subset, i.e., $f(M_{\mathbb{R}}^g)$ is a connected subset of M^g.*

This follows from the observations that real hyperelliptic curves form a connected subset of M^g. For a proof see [12].

Any orientation reversing involution $\sigma : \Sigma \to \Sigma$ of the genus g compact topological surface Σ induces an antiholomorphic involution $\sigma^* : M^g \to M^g$. Observe that the induced involution σ^* does not depend on the choice of the involution $\sigma : \Sigma \to \Sigma$. For if $\tau : \Sigma \to \Sigma$ is another involution, then

$$\tau = \tau \circ \sigma \circ \sigma = g \circ \sigma$$

where g is an orientation preserving automorphism of Σ. Hence $\tau^* = \sigma^* \circ g^* = \sigma^*$ since $g^* \in \Gamma^g = \Gamma(\Sigma)$.

It follows that

$$f(M_{\mathbb{R}}^g) \subset M_{\sigma^*}^g = \text{ the fixed-point set of } \sigma^*.$$

The complex moduli space M^g is a complex space together with the real structure σ^*. The fixed-point set $M_{\sigma^*}^g$ is the real part of M^g. The conclusion is that real curves are contained in the real part of the complex moduli space. A surprise is that there are complex curves which have real moduli but which curves are not isomorphic to real curves. Such curves have non-trivial automorphisms. This argument yields the following:

Theorem 11. *Real algebraic curves of genus g, $g > 3$, in the moduli space of complex algebraic genus g curves form the quasiregular real part of the real structure σ^* of the complex space moduli M^g.*

For a proof see [32]. The quasiregular real part consists of those real points where the local dimension of the real part is as large as possible (i.e. $3g - 3$).

These results characterize the space of smooth real algebraic curves in the moduli space of complex algebraic curves.

6.4. Moduli spaces of stable real curves.

A topology can be defined on the moduli space of smooth genus g, $g > 1$, Riemann surfaces by using the Fenchel–Nielsen coordinates (see Section 3 or [39]). To that end, fix a decomposition of a topological genus g surface into pairs of pants. Let α_j, $j = 1, 2, \ldots, 3g - 3$, be the decomposing curves. We now take a geometric point of view to moduli, and consider, instead of complex structures, the associated hyperbolic metrics. Without loss of generality we may assume that all hyperbolic metrics under consideration are such that the decomposing curves α_j are simple closed geodesic curves.

Let now $\ell_j(X)$ denote the length of the geodesic curve α_j on the Riemann surface X. Let $\theta_j(X)$ be the corresponding gluing angle (cf. Figure 1). It turns out that the parameters ℓ_j and θ_j are real analytic functions on the Teichmüller space $T(\Sigma)$ (cf. [39]). Hence the topology of the Teichmüller space (and that of the moduli space) could also have been defined by requiring the Fenchel–Nielsen parameters be continuous.

These parameters are useful when picturing degenerations of Riemann surfaces. Let $([X_k])$ be a degenerating sequence in the moduli space $M^g = M(\Sigma)$. By considerations of Section 3, we may choose, for each X_k, a decomposition of Σ into pairs of pants in such a way that the lengths of the decomposing curves are always $< 21g$. Important here is that the lengths are bounded; the bound $21g$ is of less interest.

Now there are only finitely many combinatorially different decompositions of Σ into pairs of pants. Hence, by passing to a subsequence $([X_{k_i}])$, we may assume that the lengths of the decomposing curves of a *fixed* pants decomposition of Σ are bounded.

This means that we may assume, again by passing to a subsequence, that the sequences of real numbers, formed by the Fenchel–Nielsen coordinates,

$$\ell_j(X_{k_i}) \text{ and } \theta_j(X_{k_j})$$

converge to finite limits ℓ_j^∞ and θ_j^∞. These limits specify the limiting Riemann surface. If $\ell_j^\infty = 0$ for some j, then the curve α_j gets pinched to a node. The limiting Riemann surface is a stable Riemann surface with nodes (see the exposition of Lipman Bers [3]).

The same description of degeneration applies to real algebraic curves. That is based on the fact that if X is a genus g, $g > 1$, Riemann surface admitting an antiholomorphic involution σ, then a length controlled pants decomposition can always be chosen so that it is σ invariant.

Studying the possible degenerations, one can further prove ([33]) a conjecture of Klein (1892, [19, Page 8]):

Theorem 12. *The moduli space of stable real algebraic curves of genus g, $g > 1$, is a connected and compact Hausdorff space.*

It is an interesting open problem to study, in more detail, the structure of the moduli spaces $\mathcal{M}_\mathbb{R}^g$ and of $\overline{M}_\mathbb{R}^g$, in particular to compute homology and cohomology of these spaces.

REFERENCES

[1] Lars V. Ahlfors. The complex analytic structure of the space of closed Riemann surfaces. In Rolf Nevanlinna et. al., editor, *Analytic Functions*, pages 45 – 66. Princeton University Press, 1960.

[2] Norman L. Alling and Newcomb Greenleaf. *Foundations of the theory of Klein surfaces*. Number 219 in Lecture Notes in Mathematics. Springer–Verlag, Berlin–Heidelberg–New York, 1971.

[3] Lipman Bers. Finite dimensional Teichmüller spaces and generalizations. *Bull. Amer. Math. Soc.*, 5(2):131 – 172, September 1981.

[4] Lipman Bers. An Inequality for Riemann Surfaces. In Isaac Chavel and Hersel M. Farkas, editors, *Differential Geometry and Complex Analysis*, pages 87 – 93. Springer–Verlag, Berlin–Heidelberg–New York, 1985.

[5] S. A. Broughton, E. Bujalance, A. F. Costa, J. M. Gamboa, and G. Gromadzki. Symmetries of Accola-Maclachlan and Kulkarni surfaces. *Proc. Amer. Math. Soc.*, 127(3):637–646, 1999.

[6] E. Bujalance and A. F. Costa. On symmetries of p-hyperelliptic Riemann surfaces. *Math. Ann.*, 308(1):31–45, 1997.

[7] E. Bujalance, A. F. Costa, and D. Singerman. Application of Hoare's theorem to symmetries of Riemann surfaces. *Ann. Acad. Sci. Fenn. Ser. A I Math.*, 18(2):307–322, 1993.

[8] Emilio Bujalance, Grzegorz Gromadski, and David Singerman. On the number of real curves associated to a complex algebraic curve. *Proc. Amer. Math. Soc.*, 120(2):507–513, 1994.

[9] Peter Buser and Mika Seppälä. Symmetric pants decompositions of Riemann surfaces. *Duke Math. J.*, 67(1):39 – 55, 1991.

[10] Peter Buser and Mika Seppälä. Short homology bases of Riemann surfaces. *Topology*, 2001.

[11] Peter Buser and Mika Seppälä. Real structures of Teichmüller spaces, Dehn twists, and moduli spaces of real curves. *Math Z*, 1999.

[12] Peter Buser, Mika Seppälä, and Robert Silhol. Triangulations and moduli spaces of Riemann surfaces with group actions. *Manuscripta math*, 88:209–224, 1995.

[13] C. J. Earle. On moduli of closed Riemann surfaces with symmetries. In *Advances in the Theory of Riemann Surfaces*. *Annals of Mathematics Studies 66*, pages 119 – 130, Princeton, New Yersey, 1971. Princeton University Press and University of Tokyo Press.

[14] Paola Frediani. Real algebraic functions, real algebraic curves and their moduli spaces, 1998. Thesis, University of Pisa.

[15] Frederick P. Gardiner. *Teichmüller Theory and Quadratic Differentials*. A Wiley–Interscience of Series of Texts, Monographs, and Tracts. John Wiley & Sons, New York Chichester Brisbane Toronto Singapore, 1987.

[16] J. Huisman. Real quotient singularities and nonsingular real algebraic curves in the boundary of the moduli space. *Compositio Mathematica*, 118(1):43–60, 1999.

[17] Milagros Izquierdo and David Singerman. Pairs of symmetries of Riemann surfaces. *Ann. Acad. Sci. Fenn. Math.*, 23(1):3–24, 1998.

[18] Felix Klein. Über eine neue Art von Riemannschen Flächen. *Math. Annalen*, 10, 1876.

[19] Felix Klein. Über Realitätsverhältnisse bei der einem beliebigen Geschlechte zugehörigen Normalkurve der φ. *Math. Annalen*, 42, 1892.

[20] O. Lehto. Quasiconformal homeomorphisms and Beltrami equations. In W. J. Harvey, editor, *Discrete Groups and Automorphic Functions*, pages 121 – 142. Academic Press, 1977.

[21] O. Lehto and K. I. Virtanen. *Quasiconformal mappings in the plane*, volume 126 of *Die Grundlehren der mathematischen Wissenschaften*. Springer–Verlag, Berlin–Heidelberg–New York, 1973. Translated from the German by K. W. Lucas. 2nd ed.

[22] Olli Lehto. *Univalent Functions and Teichmüller Spaces*, volume 109 of *Graduate Texts in Mathematics*. Springer-Verlag, Berlin–Heidelberg–New York, 1986.

[23] S. M. Natanzon. Automorphisms of the Riemann surface of an *M*-curve. *Funktsional. Anal. i Prilozhen.*, 12(3):82–83, 1978.

[24] S. M. Natanzon. The order of a finite group of homeomorphisms of a surface onto itself, and the number of real forms of a complex algebraic curve. *Dokl. Akad. Nauk SSSR*, 242(4):765–768, 1978.

[25] S. M. Natanzon. Real frames of complex algebraic curves, and Coxeter groups. *Uspekhi Mat. Nauk*, 51(6(312)):215–216, 1996.

[26] S.M. Natanzon. Moduli of real algebraic curves. *Uspehi Mat. Nauk*, 30:1(181):251 – 252, 1975. (Russian).

[27] Juha Pöyhönen. On Fenchel–Nielsen type coordinates for Teichmüller spaces of Klein surfaces. *Ann. Acad. Sci. Fenn. Ser. A I. Mathematica Diss.*, 72:1 – 34, 1988.

[28] John G. Ratcliffe. *Foundations of Hyperbolic Manifolds*. Springer-Verlag, New York, 1994.

[29] H. L. Royden. Automorphisms and isometries of Teichmüller space. In Cabiria Andreian Cazacu, editor, *Proceedings of the Romanian-Finnish Seminar on Teichmüller Spaces and Quasiconformal Mappings, Brasov 1969*, pages 273 – 286. Publishing House of the Academy of the Socialist Republic of Romania, 1971.

[30] H.L. Royden. Automorphisms and isometries of Teichmüller space. In Lars V. Ahlfors et al., editor, *Advances in the Theory of Riemann Surfaces*, volume 66 of *Ann. of Math. Studies*, pages 369–383. 1971.

[31] Mika Seppälä. Teichmüller Spaces of Klein Surfaces. *Ann. Acad. Sci. Fenn. Ser. A I Mathematica Dissertationes*, 15:1 – 37, 1978.

[32] Mika Seppälä. Quotients of complex manifolds and moduli spaces of Klein surfaces. *Ann. Acad. Sci. Fenn. Ser. A. I. Math.*, 6:113 – 124, 1981.

[33] Mika Seppälä. Moduli spaces of stable real algebraic curves. *Ann. scient. Éc. Norm. Sup.*, 4e série(24):519 – 544, 1991.

[34] Mika Seppälä and Robert Silhol. Moduli spaces for real algebraic curves and real abelian varieties. *Math. Z.*, 201:151 – 165, 1989.

[35] Mika Seppälä and Tuomas Sorvali. *Geometry of Riemann Surfaces and Teichmüller Spaces*. Number 169 in Mathematics Studies. North–Holland, 1992.

[36] R. J. Sibner. Symmetric fuchsian groups. *Am. J. Math.*, 90:1237–1259, 1968.

[37] Robert Silhol. Period Matrices and the Schottky Problem. In Emilio Bujalance et. al., editor, *Topics on Riemann Surfaces and Fuchsian Groups*. Cambridge University Press, 2001.

[38] Guido Weichhold. Über symmetrische Riemannsche Flächen und die Periodizitätsmodulen der zugehörigen Abelschen Normalintegrale erstes Gattung. *Leipziger Dissertation*, 1883.

[39] Scott Wolpert. The Fenchel–Nielsen deformation. *Ann. Math., II. Ser.*, 115:501–528, 1982.

[40] Li Zhong. Nonuniqueness of Geodesics in Infinite Dimensional Teichmüller spaces. *Research Report*, (29):1 – 25, 1990.

DEPARTMENT OF MATHEMATICS, FLORIDA STATE UNIVERSITY, TALLAHASSEE, FL 32306

PERIOD MATRICES AND THE SCHOTTKY PROBLEM

R. SILHOL

0. Introduction.

In this talk we will explain what is the period matrix of an algebraic curve (over $\mathbb{C}$) and prove some of its properties (Riemann's equality and Riemann's inequality). We will explain how are related period matrices of isomorphic curves and describe the action of the symplectic group on the space of period matrices. We will state (without proof) the converse of this construction which is Torelli's theorem and develop some of the consequences of this theorem.

We will then state Schottky's problem, and explain how one can check, using Theta characteristics, if a given matrix is the period matrix of a *smooth* genus 2 or 3 curve, and finally state the solution to Schottky's problem in genus 4.

1. The intersection form on a Riemann surface.

Let S be a compact Riemann surface of genus g, without boundary (or equivalently a smooth projective algebraic curve of genus g). As a topological surface S is on orientable compact surface of genus g, and is in fact oriented by the canonical orientation induced by the complex structure.

In particular, $H_1(S, \mathbb{Z}) \cong \mathbb{Z}^{2g}$ and there is a canonical perfect pairing,

$$H_1(S, \mathbb{Z}) \times H_1(S, \mathbb{Z}) \longrightarrow \mathbb{Z} \cong H_2(S, \mathbb{Z}) \ ,$$

the intersection form (see for example Farkas and Kra [Fa-Kr]). The fact that the pairing is perfect implies that the intersection form is unimodular and the fact that the dimension of S is 2 implies that this form is anti-symmetric. By classical results on bilinear forms these two facts put together mean that we can choose bases of $H_1(S, \mathbb{Z})$ in which the matrix of the intersection form is,

$$(1.1) \qquad\qquad E = \begin{pmatrix} 0_g & -1_g \\ 1_g & 0_g \end{pmatrix} \ .$$

We will say that a basis of $H_1(S, \mathbb{Z})$ is *symplectic* if in this basis the matrix of the intersection form is the one given in (1.1).

155

(1.2) Remark. 1) In the classical literature on the subject such bases are often called canonical. We have refrained from using this word because, as we will see, the choice of such a basis is far from canonical in the modern sense of the word.

2) Many authors give as standard matrix for the intersection form the transpose of the one we have given in (1.1). The motivation for our choice is that if one wants simultaneously nice properties for the period matrix (see **3**) and elegant transformation formulas — see (2.4) — one is led to the form we have given. For a more general discussion on the subject see [La-Bi], p. 213 and 322.

2. Period matrices of Riemann surfaces.

Let $\Omega^1(S)$ be the space of holomorphic differential forms on S. Then it is known since Riemann that $\Omega^1(S) \cong \mathbb{C}^g$. Let $(\omega_1, \ldots, \omega_g)$ be a basis of $\Omega^1(S)$ and let $\mathcal{B} = (\alpha_1, \ldots, \alpha_g, \beta_1, \ldots, \beta_g)$ be a **symplectic** basis of $H_1(S, \mathbb{Z})$.

Since we have 1-forms and 1-cycles we can integrate. So let,

$$X = \left(\int_{\alpha_i} \omega_j \right)_{i,j} \quad \text{and} \quad Y = \left(\int_{\beta_i} \omega_j \right)_{i,j} .$$

It turns out that both X and Y are invertible matrices, so it makes sense to define,

(2.1) Definition. *The matrix,*

$$Z = X \cdot Y^{-1} ,$$

with X and Y as above, is a period matrix of S. More accurately Z is the period matrix of S associated to the symplectic basis $\mathcal{B}$.

(2.2) Remark. The definition we have just given for the period matrix is not a universally accepted one. In general a period matrix of a Riemann surface is presented as a $g \times 2g$ (or $2g \times g$) matrix. In this classical notation the period matrix of (2.1) would be $(Z, 1_g)$. There is something important that we are hiding here (the polarization) and that will reappear later. But for the present our definition is sufficient for what we have to do.

We have made choices to define Z. We are now going to examine the influence of these choices.

The first choice we have made is the choice of a basis of $\Omega^1(S)$. Let $(\omega'_1, \ldots, \omega'_g)$ be another choice and let B be the base change matrix. Then clearly X and Y are replaced by $X \cdot B$ and $Y \cdot B$, hence the new period matrix Z' is equal to Z. In other words the period matrix is independent of the choice of the basis of $\Omega^1(S)$. This justifies the last statement of (2.1).

The other choice we have made is the choice of a symplectic basis of $H_1(S, \mathbb{Z})$. This is less innocuous. Let $(\alpha_1', \ldots, \alpha_g', \beta_1', \ldots, \beta_g')$ be another symplectic basis and let,

$$(2.3) \qquad M = \begin{pmatrix} a & b \\ c & d \end{pmatrix} \qquad \text{with } a, \ b, \ c \text{ and } d \in \operatorname{Mat}_{g \times g}(\mathbb{Z}),$$

be the base change matrix. Note that since the bases (α, β) and (α', β') are both symplectic, the matrix M of (2.3) must be in $\operatorname{Sp}_{2g}(\mathbb{Z})$, that is the subgroup of $\operatorname{GL}_{2g}(\mathbb{Z})$ consisting of matrices M such that ${}^t M E M = E$ where E is as in (1.1). Let X' and Y' be the matrices obtained with the choice (α', β'). It is easily checked that,

$$\begin{pmatrix} X' \\ Y' \end{pmatrix} = \begin{pmatrix} aX + bY \\ cX + dY \end{pmatrix}$$

and hence that, $Z' = (aX + bY)(cX + dY)^{-1} = (aX + bY)Y^{-1}Y(cX + dY)^{-1}$ and hence finally,

$$(2.4) \qquad\qquad Z' = (aZ + b)(cZ + d)^{-1}$$

(this is the elegant formula, due to Siegel, we alluded to in **1**. Note that in terms of (2.2) this imposes the order $(Z, 1_g)$, and hence (1.1), as can be easily checked by considering the case $g = 1$).

We end this section by noting that we can reformulate (2.4) by stating that if S and S' are two isomorphic Riemann surfaces and Z and Z' are period matrices of S and S' respectively, then there exists $\begin{pmatrix} a & b \\ c & d \end{pmatrix} \in \operatorname{Sp}_{2g}(\mathbb{Z})$ such that (2.4) holds.

3. Special properties of period matrices and the Siegel space.
Let Z be a period matrix of a Riemann surface as in **2**. Then Z has the following two important properties.

$$(3.1) \qquad\qquad Z = {}^t Z \quad \text{i.e, } Z \text{ is symmetric.}$$
$$(3.2) \qquad\qquad \Im m(Z) \text{ is positive definite.}$$

(3.1) is called **Riemann's equality** and (3.2) is called **Riemann's inequality**.

Both these properties are consequences of the existence of the intersection form described in **1** and we can sketch a proof as follows.

Let (α, β) be the symplectic basis of $H_1(S, \mathbb{Z})$ used to compute the period matrix Z. Let ω be a C^∞ differential 1-form and let,

$$(3.3) \qquad [\omega] = \sum_{k=1}^{g} \left(\int_{\alpha_k} \omega \right) \beta_k - \sum_{k=1}^{g} \left(\int_{\beta_k} \omega \right) \alpha_k,$$

the left hand side being considered as an element of $H_1(S,\mathbb{C})$.

Using (3.3) it is readily checked that,

$$(3.4) \qquad \int_\gamma \omega = <[\omega],\gamma> ,$$

where $<,>$ is the linear extension of the intersection form to $H_1(S,\mathbb{C})$.

But now it is well known that,

$$H_{DR}^1(S,\mathbb{C}) \times H_1(S,\mathbb{C}) \longrightarrow \mathbb{C}$$

$$(\omega,\gamma) \mapsto \int_\gamma \omega ,$$

is a perfect pairing. In other words the map defined by (3.3) is an isomorphism from $H_{DR}^1(S,\mathbb{C})$ onto $H_1(S,\mathbb{C})$ and is in fact an explicit form of Poincaré duality. Since under Poincaré duality the exterior product of differential forms corresponds to the intersection form, we also have,

$$(3.5) \qquad \int_S \omega \wedge \eta = <[\omega],[\eta]> .$$

Since the period matrix does not depend on the choice of the basis of $\Omega^1(S)$, we can always choose the basis $\{\omega_k\}$ such that $\int_{\beta_i} \omega_j = \delta_{ij}$, the Kronecker symbol. With such a choice we have $Z = \left(\int_{\alpha_i} \omega_j \right)_{i,j}$. A holomorphic 1-form is locally of the form $f(z)dz$, where z is a local coordinate and f is holomorphic. So we have, for all (i,j), $\omega_i \wedge \omega_j = 0$. Applying (3.5) to this we find,

$$\int_{\alpha_i} \omega_j - \int_{\alpha_j} \omega_i = 0 ,$$

in other words Z is symmetric, and this proves (3.1).

To obtain (3.2) write $\omega = u(z)dz$. Then, using (3.3) and (3.4), we have.

$$<[\overline{\omega}],[\omega]> = \int_{[\omega]} \overline{\omega} = \sum_{k=1}^g \left(\int_{\alpha_k} \omega \int_{\beta_k} \overline{\omega} - \int_{\beta_k} \omega \int_{\alpha_k} \overline{\omega} \right)$$

$$= \sum_{k=1}^g \left(\int_{\alpha_k} \omega \int_{\beta_k} \overline{\omega} - \overline{\int_{\beta_k} \overline{\omega} \int_{\alpha_k} \omega} \right) = 2i\Im m \left(\sum_{k=1}^g \int_{\alpha_k} \omega \int_{\beta_k} \overline{\omega} \right)$$

On the other hand using (3.5), we get,

$$<[\overline{\omega}],[\omega]> = \int_S \overline{\omega} \wedge \omega = \int_S |u(z)|^2 d\overline{z} \wedge dz$$

$$= 2i \int_S |u(z)|^2 dx \wedge dy .$$

Since $dx \wedge dy$ is a volume form it follows that,

$$\Im m \left(\sum_{k=1}^{g} \int_{\alpha_k} \omega \int_{\beta_k} \overline{\omega} \right) > 0 \ .$$

Applying this to $\omega = \sum x_k \omega_k$, it is straightforward to recover (3.2).

The basic consequence of (3.1) and (3.2) is that the period matrix Z is in the Siegel upper-half space,

$$\mathfrak{H}_g = \{ Z \in \mathrm{GL}_g(\mathbb{C}) \ / \ {}^t Z = Z, \ \Im m(Z) \text{ positive definite } \}$$

4. Torelli's theorem, the Jacobian variety and Theta functions.
A very remarkable and highly non trivial fact is that (2.4) admits a converse. This is,

(4.1) Torelli's theorem. *Let S and S' be two Riemann surfaces of genus g and let Z and Z' be period matrices of S and S' respectively. Then S and S' are isomorphic if and only if there exists a matrix $M = \begin{pmatrix} a & b \\ c & d \end{pmatrix} \in \mathrm{Sp}_{2g}(\mathbb{Z})$ such that,*

$$(4.2) \qquad\qquad Z' = (aZ + b)(cZ + d)^{-1} \ .$$

More precisely, let $\mathcal{B}$ and $\mathcal{B}'$ be the symplectic bases with which Z and Z' were computed. If Z and Z' satisfy (4.2) for some matrix $M \in \mathrm{Sp}_{2g}(\mathbb{Z})$, then there exists an isomorphism $\varphi : S \to S'$, such that the matrix of $h_1(f) : \mathrm{H}_1(S, \mathbb{Z}) \to \mathrm{H}_1(S', \mathbb{Z})$ in the bases $\mathcal{B}$ and $\mathcal{B}'$ is either M or $-M$ (both possibilities occur if and only if S is hyperelliptic).

The formulation we have given in (4.1) has the advantage of being relatively simple; it has however the disadvantage of hiding something important: the role of the polarization.

To indicate briefly what this is about let Λ be the lattice generated in $\mathbb{C}^g$ by the columns of $(Z, 1_g)$ and let $\mathrm{Jac}(S)$ be the complex torus $\mathbb{C}^g / \Lambda$. This is the *Jacobian variety* of S.

Since $\mathrm{Jac}(S)$ is a complex torus and hence a product of S^1's we can apply Künneth's formula which yields (see for example [Mu1]),

$$\bigwedge\nolimits^2 \mathrm{H}^1(\mathrm{Jac}(S), \mathbb{Z}) \cong \mathrm{H}^2(\mathrm{Jac}(S), \mathbb{Z}) \ .$$

On the other hand we have by construction canonical isomorphisms between $\mathrm{H}_1(S, \mathbb{Z})$, Λ and $\mathrm{H}_1(\mathrm{Jac}(S), \mathbb{Z})$. From these facts put together we see that $\mathrm{H}^2(\mathrm{Jac}(S), \mathbb{Z})$ can be canonically identified with the set of alternating 2-forms on $\mathrm{H}_1(S, \mathbb{Z})$. In particular the intersection form $<,>$ on S defines a class θ in $\mathrm{H}^2(\mathrm{Jac}(S), \mathbb{Z})$.

In this context (3.1) and (3.2) take a special meaning. They imply (see for example [Mu1] or [La-Bi]) that,

(4.3) *The class θ defined above is the first Chern class of an ample divisor on* $\mathrm{Jac}(S)$. *Such a class is called a polarization of* $\mathrm{Jac}(S)$.

In particular $\mathrm{Jac}(S)$ can be embedded in some projective space and hence, by a theorem of Chow, is an algebraic variety i.e, an Abelian variety.

With these notions we can reformulate Torelli's theorem,

(4.4) Torelli's Theorem (second version). *The Riemann surfaces S and S' are isomorphic if and only if there exists an isomorphism, $\varphi : \mathrm{Jac}(S) \to \mathrm{Jac}(S')$, such that $h^2(\varphi)(\theta') = \theta$, where θ and θ' are the polarizations of the jacobians associated to S and S' respectively and $h^2(\varphi)$ is the map induced by φ on the* H^2's.

We have only reformulated here the first part of (4.1), but one can of course also reformulate in this context the second part.

We have stated above that θ is the Chern class of an ample divisor. We can in fact be much more precise and explicitly construct this divisor. For this let,

$$(4.5) \qquad \Theta(Z, z) = \sum_{n \in \mathbb{Z}^g} \exp(\pi i\, {}^t nZn + 2\pi i\, {}^t nz) \ .$$

Since $\Im m(Z)$ is positive definite, the series in (4.5) is uniformly convergent on compact subsets of $\mathbb{C}^g$ and hence, $z \mapsto \Theta(Z, z)$ defines a holomorphic function on $\mathbb{C}^g$.

We have obviously, for $m \in \mathbb{Z}^g$,

$$(4.6) \qquad\qquad \Theta(Z, z + m) = \Theta(Z, z)$$

and also, since Z is symmetric,

(4.7)
$$\Theta(Z, z + Zm)$$
$$= \sum_{n \in \mathbb{Z}^g} \exp(\pi i(\,{}^t nZn + {}^t nZm + {}^t mZn + 2\,{}^t nz))$$
$$= \sum_{n \in \mathbb{Z}^g} \exp(\pi i(\,{}^t(n + m)Z(n + m) + 2\,{}^t(n + m)z - {}^t mZm - 2\,{}^t mz))$$
$$= \exp(-\pi i\,{}^t mZm - 2\pi i\,{}^t mz)\Theta(Z, z) \ .$$

Let $\mathcal{Z}(\Theta)$ be the set of zeros of $\Theta(Z, z)$ in $\mathbb{C}^g$ and let, as above, Λ be the lattice generated by the columns of $(Z, 1_g)$. Then, (4.6) and (4.7) can be reformulated by saying that $\mathcal{Z}(\Theta)$ is globally invariant under the action of Λ or otherwise said that $\mathcal{Z}(\Theta)$ defines a complex codimension 1 subvariety of $\mathrm{Jac}(S)$ (it can be easily proved that it is non empty), i.e, a divisor. We will denote again $\mathcal{Z}(\Theta)$ this divisor. The result is that $c_1(\mathcal{Z}(\Theta)) = \theta$, where c_1 denotes the first Chern class and θ is the above defined polarization.

5. The Schottky problem.

A classical result, already known to Riemann, states that the space of curves of genus $g > 1$ is of dimension $3g - 3$. On the other hand the Siegel upper half space $\mathfrak{H}_g$ is of dimension $\frac{g(g+1)}{2}$. In other words for $g \geqslant 4$ the period map cannot be surjective. In fact we are also going to indicate why even for $g = 2$ and 3 this map is not surjective.

The problem of finding a criterion for a given matrix $Z \in \mathfrak{H}_g$ to be the period matrix of a Riemann surface of genus g is called the Schottky problem.

We first look at the problem in the genus 2 case. Let $Z \in \mathfrak{H}_2$ and let Λ, Θ and $\mathcal{Z}(\Theta)$ be as before. Let $A = \mathbb{C}^g / \Lambda$.

For simplicity we write D for the divisor $\mathcal{Z}(\Theta)$ on A. Looking at the behaviour of D under translations one can prove that the self-intersection $(D \cdot D) = 2$. Using the adjunction formula and the fact that an abelian variety cannot contain a rational curve, it is possible from this to conclude that either,

    (1) D is a smooth genus 2 algebraic curve.

or,

    (2) D is the union of two elliptic curves E_1 and E_2 intersecting in one point.

In the first case A is the jacobian of D and D defines the polarization. In the second case A is isomorphic to the product of the two elliptic curves and it can be proved that A together with the polarization defined by D cannot be the polarized jacobian of a genus 2 Riemann surface.

Stated otherwise this means that $Z \in \mathfrak{H}_2$ is the period matrix of a genus 2 Riemann surface (or a smooth projective genus 2 curve) if and only if Z is not in the orbit of $\left\{ \begin{pmatrix} \tau_1 & 0 \\ 0 & \tau_2 \end{pmatrix} \ / \ \mathfrak{Im}(\tau_i) > 0 \right\}$ under the action of $\mathrm{Sp}_4(\mathbb{Z})$.

(5.1) Remark. A word of caution should be given here. As an abelian variety (forgetting the polarization) the product of two elliptic curves can be isomorphic to the jacobian of a smooth genus 2 curve (see for example [Ek-Se]). The above statement is that $E_1 \times E_2$ together with the polarization $(E_1 \times \{p_2\}) \cup (\{p_1\} \times E_2)$ cannot be the polarized jacobian of a smooth genus 2 curve.

For further generalizations we are going to give another characterization. We introduce the following notation,

(5.2)

$$\Theta \begin{bmatrix} 2\alpha \\ 2\beta \end{bmatrix} (Z) = \exp(\pi i \,^t\alpha Z\alpha + 2\pi i \,^t\alpha\beta) \; \Theta(Z, Z\alpha + \beta)$$

$$= \sum_{n \in \mathbb{Z}^g} \exp(\pi i \,^t(n + \alpha)Z(n + \alpha) + 2\pi i \,^t(n + \alpha)\beta) \,,$$

2α and 2β in $\mathbb{Z}^g$.

For a and b in $\mathbb{Z}^g$ we will say that $\begin{bmatrix} a \\ b \end{bmatrix}$ is an even (respectively odd) characteristic if ${}^t a \cdot b$ is even (respectively odd). If $\begin{bmatrix} a \\ b \end{bmatrix}$ is odd then it is readily checked that $\Theta \begin{bmatrix} a \\ b \end{bmatrix} (Z) = 0$. On the other hand the cancelation of $\Theta \begin{bmatrix} a \\ b \end{bmatrix} (Z)$ for an even characteristic corresponds to non trivial facts (see for example Mumford [Mu3] Vol. II). In particular if Z is the period matrix of a genus 2 Riemann surface then $\Theta \begin{bmatrix} a \\ b \end{bmatrix} (Z) \neq 0$ for all even characteristics. But now,

$$(5.3) \qquad \Theta \begin{bmatrix} a_1, a_2 \\ b_1, b_2 \end{bmatrix} \begin{pmatrix} \tau_1 & 0 \\ 0 & \tau_2 \end{pmatrix} = \Theta \begin{bmatrix} a_1 \\ b_1 \end{bmatrix} (\tau_1) \cdot \Theta \begin{bmatrix} a_2 \\ b_2 \end{bmatrix} (\tau_2)$$

(where for typographic reasons we have written a and b as line vectors).

In particular we will have,

$$\Theta \begin{bmatrix} 1 & 1 \\ 1 & 1 \end{bmatrix} \begin{pmatrix} \tau_1 & 0 \\ 0 & \tau_2 \end{pmatrix} = 0 \ .$$

Replacing Z by a transform under $\mathrm{Sp}_{2g}(\mathbb{Z})$ does not change the number of zero theta characteristics and does not change the character, even or odd, of these characteristics. Reformulating we have,

(5.4) *A matrix Z in $\mathfrak{H}_2$ is the period matrix of a genus 2 Riemann surface if and only if, $\Theta \begin{bmatrix} a \\ b \end{bmatrix} (Z) \neq 0$, for all even characteristics $\begin{bmatrix} a \\ b \end{bmatrix}$.*

For similar reasons this generalizes to genus 3. We have, $Z \in \mathfrak{H}_3$ is the period matrix of a genus 3 Riemann surface if and only if Z is not in the orbit of $\{ \begin{pmatrix} \tau_1 & 0 \\ 0 & W \end{pmatrix} \ / \ \tau_1 \in \mathfrak{H}_1, \ W \in \mathfrak{H}_2 \}$, under the action of $\mathrm{Sp}_6(\mathbb{Z})$.

For matrices of the form $\begin{pmatrix} \tau_1 & 0 \\ 0 & W \end{pmatrix}$ we have the decomposition,

$$\Theta \begin{bmatrix} a_1, a_2, a_3 \\ b_1, b_2, b_3 \end{bmatrix} \begin{pmatrix} \tau_1 & 0 \\ 0 & W \end{pmatrix} = \Theta \begin{bmatrix} a_1 \\ a_2 \end{bmatrix} (\tau_1) \cdot \Theta \begin{bmatrix} a_2, a_3 \\ b_2, b_3 \end{bmatrix} (W) \ .$$

This leads to,

(5.5) *A matrix Z in $\mathfrak{H}_3$ is the period matrix of a genus 3 Riemann surface if and only if $\Theta \begin{bmatrix} a \\ b \end{bmatrix} (Z) = 0$ for at most one even characteristic $\begin{bmatrix} a \\ b \end{bmatrix}$, and if one is zero then the Riemann surface is hyperelliptic.*

For $g = 4$, the set of period matrices of algebraic curves of genus 4 (possibly degenerated) is of codimension 1 in $\mathfrak{H}_4$. In 1888 Schottky exhibited a polynomial of degree 16 in the Theta characteristics, identically zero on the set of period matrices of curves, but not on $\mathfrak{H}_4$. It was only proved relatively recently, by Igusa [Ig], that this relation indeed completely characterizes such matrices.

To express this polynomial, write,

$$R_1 = \Theta\begin{bmatrix} 0000 \\ 0000 \end{bmatrix} \Theta\begin{bmatrix} 0000 \\ 1000 \end{bmatrix} \Theta\begin{bmatrix} 0000 \\ 0100 \end{bmatrix} \Theta\begin{bmatrix} 0000 \\ 1100 \end{bmatrix} \Theta\begin{bmatrix} 0000 \\ 0010 \end{bmatrix} \Theta\begin{bmatrix} 0000 \\ 1010 \end{bmatrix} \Theta\begin{bmatrix} 0000 \\ 0110 \end{bmatrix} \Theta\begin{bmatrix} 0000 \\ 1110 \end{bmatrix}$$

$$R_2 = \Theta\begin{bmatrix} 0000 \\ 0001 \end{bmatrix} \Theta\begin{bmatrix} 0000 \\ 1001 \end{bmatrix} \Theta\begin{bmatrix} 0000 \\ 0101 \end{bmatrix} \Theta\begin{bmatrix} 0000 \\ 1101 \end{bmatrix} \Theta\begin{bmatrix} 0000 \\ 0011 \end{bmatrix} \Theta\begin{bmatrix} 0000 \\ 1011 \end{bmatrix} \Theta\begin{bmatrix} 0000 \\ 0111 \end{bmatrix} \Theta\begin{bmatrix} 0000 \\ 1111 \end{bmatrix}$$

$$R_3 = \Theta\begin{bmatrix} 0001 \\ 0000 \end{bmatrix} \Theta\begin{bmatrix} 0001 \\ 1000 \end{bmatrix} \Theta\begin{bmatrix} 0001 \\ 0100 \end{bmatrix} \Theta\begin{bmatrix} 0001 \\ 1100 \end{bmatrix} \Theta\begin{bmatrix} 0001 \\ 0010 \end{bmatrix} \Theta\begin{bmatrix} 0001 \\ 1010 \end{bmatrix} \Theta\begin{bmatrix} 0001 \\ 0110 \end{bmatrix} \Theta\begin{bmatrix} 0001 \\ 1110 \end{bmatrix}$$

(where to simplify we have written $\Theta\begin{bmatrix} a \\ b \end{bmatrix}$ in place of $\Theta\begin{bmatrix} a \\ b \end{bmatrix}(Z)$). The polynomial,

$$R_1^2 + R_2^2 + R_3^2 - 2R_1R_2 - 2R_1R_3 - 2R_2R_3 \ ,$$

is then invariant under the action of $\mathrm{Sp}_8(\mathbb{Z})$ and its zero set is the closure of the set of period matrices of genus 4 curves (see [Ig]).

For $g \geqslant 5$ there are no known criteria of this type. For other approaches see the survey paper of Beauville [Be].

REFERENCES

[Be]   A. Beauville, *Le problème de Schottky et la conjecture de Novikov*, Séminaire Nicolas Bourbaki **1986-87** (Février 1987), 675-01–12.

[Ek-Se] T. Ekedahl and J.-P. Serre, *Exemples de courbes algébriques a Jacobienne complètement décomposable*, C. R. Acad. Sci. Paris **317, série I** (1993), 509–513.

[Fa-Kr] H. Farkas and I. Kra, *Riemann Surfaces*, Springer G.T.M. 71, Berlin Heidelberg New York, 1980.

[Ig]   J. Igusa, *On the irreducibility of Schottky's divisor*, J. Fac. Sci. Tokyo **28** (1981), 531–545.

[La-Bi] H. Lange and Ch. Birkenhake, *Complex Abelian Varieties*, Springer, Berlin, Göttingen, Heidelberg, 1992.

[Mu1]   D. Mumford, *Abelian Varieties*, Oxford Univ. Press, 1970.

[Mu2]   D. Mumford, *Curves and their Jacobians*, Univ. of Michigan Press, Ann Arbor, 1975.

[Mu3]   D. Mumford, *Tata Lectures on Theta, Vol.I and II*, Birkhäuser, Boston Bassel Berlin, 1983 and 1984.

Université Montpellier II, Département de Mathématiques,
UMR CNRS 5030,
Place E. Bataillon 34095 Montpellier Cedex 5, France
rs@darboux.math.univ-montp2.fr

HURWITZ SPACES

S. M. NATANZON

Riemann surfaces P first appear in works of Riemann, as domains of definitions of complex analytic functions f. Soon after that Hurwitz [7] and others began a systematic investigation of spaces of such pairs (P, f). Now they are called Hurwitz spaces and play an important role in algebraic geometry and mathematical physics.

I recall some definitions. Let P and $\widetilde{P}$ be compact Riemann surfaces and $f : P \to \widetilde{P}$ be a mapping. Consider the points $p \in P$ and $\widetilde{p} = f(p) \in \widetilde{P}$. Let $p \in U \subset P$ and $\widetilde{p} \in \widetilde{U} \subset \widetilde{P}$ be neigborhoods such that $f(U) \subset \widetilde{U}$. Consider biholomorphic local maps $z_p : U \to \Lambda$, $\widetilde{z}_{\widetilde{p}} : \widetilde{u} \to \Lambda$, where $\Lambda = \{z \in \mathbb{C} | |z| < 1\}$, $\mathbb{C}$ is the complex plane and $z_p(p) = \widetilde{z}_{\widetilde{p}}(\widetilde{p}) = 0$. In these local coordinates the mapping f has the form $\widetilde{z}_{\widetilde{p}} = f_p(z_p)$, where $f_p : V \to \widetilde{V}$ and

$$V = z_p(U) \subset \mathbb{C}, \quad \widetilde{V} = \widetilde{z}_{\widetilde{p}}(U) \subset \mathbb{C} \quad f_p = \widetilde{z}_{\widetilde{p}} f z_p^{-1}.$$

The mapping f is called *holomorphic in p* if f_p is holomorphic that is

$$f_p(z_p) = z_p^{n_p} + a_1 z_p^{n_p+1} + a_2 z_p^{n_p+2} + \dots$$

The number n_p is called the *degree f in the point p*. If $n_p > 1$ then we say that p is a *ramification point* of f.

The mapping f is called *holomorphic* if it is holomorphic in alls points $p \in P$. In this case f has only finite number of ramification points.

Consider now $\widetilde{p} \in \widetilde{P}$ and the preimage

$$f^{-1}(\widetilde{p}) = \{p_1, ..., p_k\} \subset P.$$

The set of its degrees $\{n_{p_1}, ..., n_{p_k}\}$ is called the *type* of $\widetilde{p}$. The sum $\sum_{i=1}^{k} n_{p_i}$ does not depend on point $\widetilde{p} \in \widetilde{P}$ and is called a *degree of f*. If some $n_{p_i} > 1$ then we say that $\widetilde{p}$ is a *branch* point. Thus a branch point is the image of a ramification point.

If $\widetilde{P}$ is the Riemann sphere $\widetilde{\mathbb{C}} = \mathbb{C} \cup \infty$, then we say that the holomorphic mapping $f : \widetilde{P} \to P$ is a *meromorphic function*. In this case the genus $g(P)$

This work was carried out with the financial support of the grants: RFBR-96-15-96043, INTAS-96-0713 and supported by DGICYT of P is called genus of f. The points $f^{-1}(\infty)$ are called *poles* of f. If all finite branch points $z \in \mathbb{C}$ have type $(2, 1, 1, ..., 1)$ then we say that f is a *function in general position*.

Examples

1. Consider a polynomial $f(z) = z^n + a_1 z^{n-1} + ... + a_n$. Then f defines a mapping $f : \mathbb{C} \to \mathbb{C}$. Extend f to $\bar{f} : \bar{\mathbb{C}} \to \bar{\mathbb{C}}$ setting $\bar{f}(\infty) = \infty$. Then $\bar{f}$ is a meromorphic function of genus 0 and degree n. If $\{a_i\} \in R$ then its real part has the following form

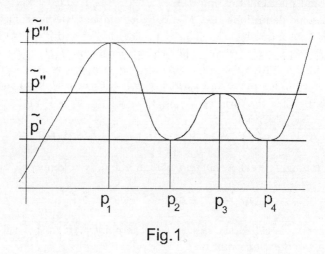

Fig.1

The real ramification points are p_1, p_2, p_3, p_4 and ∞. The real branch points are $\tilde{p}', \tilde{p}'', \tilde{p}'''$ and ∞. The types of $\tilde{p}''$ and $\tilde{p}'''$ are $(2, 1, 1, ..., 1)$. The type of $\tilde{p}'$ is $(2, 2, 1, 1, ..., 1)$. The type of ∞ is (n).

2. Consider a Laurent'polynomial $f : \mathbb{C} \setminus 0 \to \mathbb{C}$

$$f(z) = z^{-m} + b_1 z^{-m+1} + ... + b_m + a_1 z + a_2 z^2 + ... + a_{n-1} z^{n-1} + z^n.$$

Extend f to $\bar{f} : \bar{\mathbb{C}} \to \bar{\mathbb{C}}$, setting $\bar{f}(0) = \bar{f}(\infty) = \infty$. Then $\bar{f}$ is a meromorphic function of genus 0 and degree $m + n$. If $\{a_i\}, \{b_i\} \in R$, then its real part has the following form

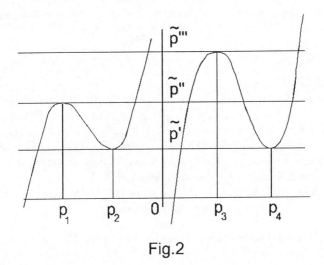

$\widetilde{p}'''$

$\widetilde{p}''$

$\widetilde{p}'$

p_1 p_2 0 p_3 p_4

Fig.2

The real ramification points are $p_1, p_2, p_3, p_4, 0$ and ∞. The real branch points are $\widetilde{p}', \widetilde{p}'', \widetilde{p}'''$ and ∞. The types of $\widetilde{p}''$ and $\widetilde{p}'''$ are $(2, 1, 1, ..., 1)$. The type of $\widetilde{p}'$ is $(2, 2, 1, 1, ..., 1)$. The type of ∞ is $(n, m, 1, 1, ..., 1)$.

3. Let G be a group of automorphisms of a Riemann surface P and $\widetilde{P} = P/G$. Then the natural projection $f : \widetilde{P} \to P$ is a holomorphic mapping and a meromorphic function if $P = \bar{\mathbb{C}}$.

4. If the meromorphic function $f : P \to \bar{\mathbb{C}}$ has only 3 branch points then f is called Belyi's function. Such functions exist if and only if P is a rational curve. Thus an investigation of such meromorphic functions gives important arithmetic information.

We will consider that the meromorphic functions $f : P \to \bar{\mathbb{C}}$ and $f' : P' \to \bar{\mathbb{C}}$ are the same if there exists a biholomorphic map $\varphi : P \to P'$ such that $f = f'\varphi$. Consider the set $H_{g,n}$ of all meromorphic functions $f : P \to \bar{\mathbb{C}}$ of genus g and degree n. The branch points of f give local coordinates on $H_{g,n}$. The space $H_{g,n}$ and its natural subspaces are called *Hurwitz spaces*.

Examples

1. The set A_n of all polynomials

$$f(z) = z^n + a_1 z^{n-1} + ... + a_n$$

in general position up to changes $z \mapsto az + b$ forms a Hurwitz space.

2. The set $L_{n,m}$ of all Laurent's polynomial

$$f(z) = z^{-m} + b_1 z^{-m+1} + ... + a_n z^n, \quad a_n \neq 0$$

in general position up to changes $z \mapsto az$ forms a Hurwitz space.

3. All functions in general position with a fixed type of poles

$$H_{g,n}(n_1, ..., n_k) =$$

$$\{f \in H_{g,n} | \infty \in \mathbb{C} \text{ is a branch point of } f \text{ of type } (n_1, ..., n_k) \}$$

form a Hurwitz space.

The spaces $H_{g,n}(n_1, ..., n_k)$ are very intresting. In this case $H_{0,n}(n) = A_n$, $H_{0,n+m}(n, m) = L_{n,m}$, and thus $H_{g,n}(n_1, ..., n_k)$ is a generalization of polynomials. Furthemore, a closure of $H_{g,n}(1, ..., 1)$ contains all $H_{g,n}$. Some years ago Dubrovin [3] showed that the spaces $H_{g,n}(n_1, ..., n_k)$ have important physical applications: they have a natural Frobenius structure and thus they are moduli spaces for some topological field theories.

It is obvious that A_n and $L_{n,m}$ are connected. In 1891 Hurwitz [7] proved that $H_{g,n}(1, 1, ..., 1)$ is connected. Furthemore

Theorem 1 (Natanzon 1984 [9, 11]). *Any space $H_{g,n}(n_1, ..., n_k)$ is connected.*

Consider now some more general type of Hurwitz spaces. We say that meromorphic functions (P_1, f_1) and (P_2, f_2) are topologically equivalent if there exist homeomorphisms $\varphi : P_1 \to P_2$ and $\psi : \bar{\mathbb{C}} \to \bar{\mathbb{C}}$ such that $\psi f_1 = f_2 \varphi$. A class t of topologocal equivalence is called a topological type.

Let H^t be the Hurwitz space of all meromorphic functions of topological type t. It follows from Theorem 2 that any $H_{g,n}(n_1, ..., n_k)$ is a space of type H^t. If $f \in H^t$ and $\alpha \in \text{Aut}(\bar{\mathbb{C}}) \cong \text{PSL}(2, \mathbb{C})$ then $\alpha f \in H^t$. Thus $\text{PSL}(2, \mathbb{C})$ acts on H^t. Let us $\widetilde{H}^t = H^t/\text{PSL}(2, \mathbb{C})$.

Theorem 2 (Natanzon 1997 [12]). *Any space $\widetilde{H}^t$ is homeomorphic to R^m/Mod, where Mod acts discretely on R^n.*

For the space of Laurent's polynomials $L_{n,m}$ it was proved (Arnold 1996 [1]) using deep theorems of theory of singularities. For the space $H_{g,n}(1, ..., 1)$ it was proved (Natanzon 1986, [10]) using the theory of Fuchsian groups. The last method is good and applies for arbitrary H^t.

This method gives also a description of the group Mod as a subgroup of finite index of a spherical braid group B_k. Thus we have the reduction of the problem of topological classification of meromorphic functions to an algebraic problem of description of subgroups of finite index in B_k. This index is the degree of the map $H^t \to (\bar{\mathbb{C}})^k$, which associates with $f \in H^t$ the branch points with $f \in H^t$. Recent works (for example [5]) show that these degrees describe important mathematical physics modeles. In appendix the author gives a proof of Theorem 2.

The compactifications of the Hurwitz spaces are specially important for mathematical and physical applications. There exist several compactifications of the Hurwitz space $H_{g,n}$. For example, Harris and Mumford compactification [6] plays an important role in algebraic geometry. Recently

Turaev and the author are proposed a purely topological compactification of $H_{g,n}$.

It is constructed by *decorated functions* $(f, E, \{D_e\}_{e \in E})$ where f is a meromorphic function from $H_{g,n}(1, ..., 1)$, $E \subset \bar{\mathbb{C}}$ is a finite subset and $D_e \subset \bar{C}$ is a disk, containing e and more than 1 branch point. Decorated functions $(f, E, \{D_e\}_{e \in E})$ and $(f', E', \{D_{e'}\}_{e' \in E})$ are equivalent if $E' = E$ and one goes to the other by a homotopy fixing $E = E'$. The compactification $N_{g,n}$ of $H_{g,n}$ is a sum of $H_{g,n}$ and equivalent classes of decorated functions. The topology of $H_{g,n}$ gives a topology on $N_{g,n}$.

Theorem 3 (Turaev, Natanzon 1999 [13]). *$N_{g,n}$ is compact and Hausdorff.*

The compactification $N_{g,n}$ gives compactifications for all other Hurwitz spaces. Our construction gives also a natural stratification $N_{g,n}$ on H^t. It follows from the topological description of $N_{g,n}$ because Theorem 2 gives a description of H^t.

Specially important is the Euler characteristic of compactifications of Hurwitz spaces. Last investigations show that these Euler characteristics define a partition function for a modern variant of unit field theory [2].

For a calculation of Euler characteristic of $N_{g,n}$ we consider regular labelled 2-coloured graphs. Such graph is a finite connected graph G such that every vertex of G is provided with one of two colours (red or blue) so that the endpoints of any edge have different colours. Every edge $\ell \in G$ is provided with a positive integer $n(\ell)$. Every vertex $\nu \in G$ is provided with a non-negative integer $g(\nu)$, where $g(\nu) = 0$, if ν is the end of only one edge ℓ and also $n(\ell) = 1$.

Let us define

$$n(G) = \sum_{\ell \in G} n(\ell), \quad g(G) = \sum_{\nu \in G} g(\nu) + 1 - \chi(G),$$

where $\chi(G)$ is the Euler characteristic of G.

Let us consider the number $\Gamma_{g,n}$ of all regular labelled 2-coloured graphs, G such that $g(G) = g$ and $n(G) = n$ and

$$n(G) + d(G) + \sum_{\nu \text{ is red}} (2g(\nu) - 2) \geqslant 2, \quad n(G) + d(G) + \sum_{\nu \text{ is blue}} (2g(\nu) - 2) \geqslant 2,$$

where $d(G)$ is the number of edges of G. Let $\chi(N_{g,n})$ be the Euler characteristic of $N_{g,n}$.

Theorem 4 (Turaev, Natanzon 1999 [13]). $\chi(N_{g,n}) = \Gamma_{g,n} + 4.$

Appendix. Proof of Theorem 2.

Let

$$\Lambda = \{z \in C \mid Im \ z > 0\}$$

be the Lobachevsky's plane. The group of its automorphisms $Aut(\Lambda)$ consists of maps

$$z \mapsto \frac{az + b}{cz + d},$$

where $a, b, c, d \in R$ and $ad - bc > 0$. An automorphism is called *elliptic* if it has one fixed point in Λ. An automorphism is called *parabolic* if it has one fixed point in R. Any automorphism is called *hyperbolic* if it has two fixed points in R. The set $Aut(\Lambda)\backslash\{id\}$ is divided on three subsets:

a set $Aut_+(\Lambda)$ of elliptic automorphisms;

a set $Aut_0(\Lambda)$ of parabolic automorphisms;

a set $Aut_-(\Lambda)$ of hyperbolic automorphisms.

Each $C \in Aut_0(\Lambda)$ is of the form

$$C(z) = \frac{(1 - a\gamma)z + a^2\gamma}{-\gamma z + (1 + a\gamma)},$$

where $C(a) = a$.

It is called *positive* if

$$\frac{1 - a\gamma}{-\gamma} < a$$

and *negative* if

$$\frac{1 - a\gamma}{-\gamma} > a.$$

Each $C \in Aut_-(\Lambda)$ is of the form

$$C(z) = \frac{(\lambda\alpha - \beta)z + (1 - \lambda)\alpha\beta}{(\lambda - 1)z + (\alpha - \lambda\beta)},$$

where $\lambda > 1$, $C(\alpha) = \alpha$, $C(\beta) = \beta$. It is called *positive* if $\alpha < \beta$ and *negative* if $\alpha > \beta$.

Let

$$C_1, C_2 \in Aut_0(\Lambda) \cup Aut_-(\Lambda).$$

We say that $C_1 < C_2$ if all fixed points of C_1 are less that all fixed points of C_2.

The set

$$\{C_1, C_2, C_3\} \in Aut_0(\Lambda) \cup Aut_-(\Lambda)$$

is called *sequential* if

$$C_1 C_2 C_3 = 1$$

and there exists

$$D \in Aut(\Lambda)$$

such that

$$\tilde{C}_i = DC_iD^{-1}$$

are positive and

$$\tilde{C}_1 < \tilde{C}_2 < \tilde{C}_3.$$

<u>Lemma 1.</u> *Let*

$$C_1 z = \lambda z \quad (\lambda > 1), \quad C_2(z) = \frac{(1 - a\gamma)z + a^2\gamma}{-\gamma z + (1 + a\gamma)} \quad and \quad C_3 = (C_1C_2)^{-1}.$$

Then $\{C_1, C_2, C_3\}$ *is sequential if and only if*

$$a > 0 \quad and \quad a\gamma \geq \frac{\sqrt{\lambda} + 1}{\sqrt{\lambda} - 1}.$$

Further

$$C_3 \in Aut_0(\Lambda)$$

if and only if

$$a\gamma = \frac{\sqrt{\lambda} + 1}{\sqrt{\lambda} - 1}.$$

<u>Proof.</u> We have

$$C_3^{-1}z = C_1C_2z = \frac{(1 - a\gamma)\lambda z + a^2\gamma\lambda}{-\gamma z + (1 + a\gamma)}.$$

Fixed points of C_3 are the solutions of the equation

$$\gamma x^2 + (\lambda - a\gamma\lambda - 1 - a\gamma)x + a^2\gamma\lambda = 0. \tag{1}$$

Thus the condition

$$C_3 \in Aut_0(\Lambda) \cup Aut_-(\Lambda)$$

is equivalent to

$$a\gamma \notin \left(\frac{\sqrt{\lambda} - 1}{\sqrt{\lambda} + 1}, \frac{\sqrt{\lambda} + 1}{\sqrt{\lambda} - 1}\right).$$

Further

$$C_3 \in Aut_0(\Lambda)$$

if and only if

$$a\gamma = \frac{\sqrt{\lambda} - 1}{\sqrt{\lambda} + 1} \quad or \quad \gamma = \frac{\sqrt{\lambda} + 1}{\sqrt{\lambda} - 1}.$$

Suppose that

$$\{C_1, C_2, C_3\}$$

is sequential. Then $a > 0$, $\gamma > 0$ and the roots of (1) are greater than a.
Thus

$$\frac{a\gamma\lambda + a\gamma + 1 - \lambda}{\gamma} > a \quad \text{and} \quad a\gamma > \frac{\lambda - 1}{\lambda} > \frac{\sqrt{\lambda} - 1}{\sqrt{\lambda} + 1}.$$

Therefore

$$a\gamma \geq \frac{\sqrt{\lambda} + 1}{\sqrt{\lambda} - 1}.$$

Conversely, suppose that

$$a > 0, \quad a\gamma \geq \frac{\sqrt{\lambda} + 1}{\sqrt{\lambda} - 1};$$

then

$$\gamma > 0, \quad \frac{1 - a\gamma}{-\gamma} < a$$

and therefore

$$C_3 > C_2.$$

Finally

$$C_3^{-1}(\infty) = \frac{1 - a\gamma}{-\gamma} > 0$$

and thus C_3 is positive. $\square$

The set

$$\{C_1, ..., C_n\}$$

is called *sequential* if the set

$$\{C_1 \cdots C_{i-1}, C_i, C_{i+1} \cdots C_n\}$$

is sequential for $i = 2, \ldots, n - 1$.
   It follows from [8].
   <u>Lemma 2</u>.   *1) Any sequential set*

$$\{C_1, ..., C_n\}$$

generates a free Fuchsian group

$$\Gamma \subset Aut_0(\Lambda) \cup Aut_-(\Lambda) \cup \{id\}$$

such that

$$\Lambda/\Gamma$$

is a sphere with k_0 punctures and k_- holes, where k_0 (respectively k_-) is a number of parabolic (respectively hyperbolic) transformations among C_i.

2) Any sphere with k_0 punctures and k_- holes is biholomorphic equivalent to Λ/Γ, where Γ is the Fuchsian group genereted by the sequential set

$$\{C_1, ..., C_n\}.$$

Denote by (S_k, δ_k) any pair, where S_k is the free group with free generators $c_1, \ldots, c_{k-1}$ and $\delta_k = \{c_1, \ldots, c_{k-1}, c_k\}$, where $c_k = (c_1 \cdots c_{k-1})^{-1}$.

Any monomorphism $\psi : S_k \to Aut(\Lambda)$ is called a *realization* of (S_k, δ_k) if $\{\psi(c_1), \ldots, \psi(c_k)\}$ is a sequential set of parabolic elements.

Let T_k^* be a set of all realizations

$$(S_k, \delta_k).$$

The group $Aut(\Lambda)$ acts in T_k^* by conjugation

$$\psi(c_i) \mapsto \gamma\psi(c_i)\gamma^{-1} \quad (\gamma \in Aut(\Lambda)).$$

Put $T_k = T_k^*/Aut(\Lambda)$. There exists a natural bijection between T_k and

$$T_k' = \{\psi \in T_k^* \mid \psi(c_1 c_2)^{-1} z = \lambda z, \quad \lambda > 1, \quad \psi(c_2)(-1) = -1\}.$$

We note that $T_k \cong T_k'$ is embedded in R^{2k-6} by a map

$$\psi \mapsto \{\lambda, \gamma_3, a_3, \gamma_4, a_4, \ldots, a_{k-2}, \gamma_{k-1}\} \in R^{2k-6},$$

where

$$\psi(c_i)z = \frac{(1 - a_i\gamma_i)z + a_i^2\gamma_i}{-\gamma_i z + (1 + a_i\gamma_i)}.$$

<u>Theorem A.</u>

$$T_k \cong R^{2k-6}.$$

<u>Proof:</u> Lemma 1 gives a description for

$$\psi \in T_k'.$$

These conditions have the form $\lambda > 1, \quad a_i > b_i$ and $a_i\gamma_i > v_i$, where

$$b_i = b_i(\lambda, \gamma_3, a_3, \ldots, \gamma_{i-1}, a_{i-1}), \quad v_i = v_i(\lambda, \gamma_3, a_3, \ldots, \gamma_{i-1}, a_{i-1}).$$

This proves the theorem. $\square$

Let M_k be the moduli space (i.e. the space of classes of biholomorphic equivalents) of spheres with $k > 3$ punctures.

<u>Theorem B</u> [4]. $M_k = T_k/Mod_k \cong R^{2k-6}/Mod_k$, where Mod_k acts discretely on T_k.

<u>Proof:</u> Let

$$\Psi^* : T_k^* \to M_k$$

be the map such that $\Psi^*(\psi) = \Lambda/\psi(S_k)$ It gives $\Psi : T_k \to M_k$. From the lemma 2 it follows that Ψ is surjective.

Moreover $\Psi(\psi_1) = \Psi(\psi_2)$ if and only if there exists a $\gamma \in Aut(\Lambda)$, a permutation $\sigma : \{1,\ldots,k\} \to \{1,\ldots,k\}$ and $\alpha_i \in S_k$ $(i = 1,\ldots,k)$ such that $\psi_1(S_k) = \gamma\psi_2(S_k)\gamma^{-1}$ and $\psi_1(c_i) = \gamma\psi_2(\alpha_i c_{\sigma(i)}\alpha_i^{-1})\gamma^{-1}$. The sets (γ,σ,α_i) form a group B_k^*. It acts on T_k and $M_k = T_k/B_k^*$. The sets $(\gamma,1,1,\ldots)$ form a normal subgroup $B_k^0 \subset B_k^*$, which is the kernel of the action of B_k^*. Thus $B_k = B_k^*/B_k^0$ acts on T_k and $M_k = T_k/B_k$.

Let $\psi \in T_k$. The metric on Λ determines a metric on $P = \Lambda/\psi(S_k)$ with a constant negative curvature. Each $c \in S_k$ gives a unique geodesic curve $\tilde{c} \subset P$. For $P \subset M_k$ in general position, the lengths $\rho(c)$ are different for different c. The system of generators allows us to enumerate all elements $c \in S_k$ and thus we have a correspondence

$$P \mapsto v = (\rho(c_1), \rho(c_2), \ldots) \in R^\infty.$$

The coordinates v form a discrete set in R and thus B_k acts discretely with respect to any sensible topology in R^∞. Moreover $\rho(c_i)$ is algebraically expressed in the coordinates $(\lambda, \gamma_3, a_3, \ldots, \gamma_{k-1})$ on T_k. Thus B_k acts discretely on T_k in our topology. According to Theorem A $T_k \cong R^{2k-6}$.

Let (P, f) be a meromorphic function. The genus g of P is called the *genus* of (P, f) and the number of sheets of f is called *degree* of (P, f). A point $p \in P$ is called a *singular point* if $df(p) = 0$. The image $f(p) \in \bar{C}$ of a singular point is called a *singular value*.

Let $a_1, \ldots, a_k \in \bar{C}$ be singular values of f. Let $b_i^1, \ldots, b_i^{m_i}$ be singular points, corresponding to a_i. Let n_i^s be the number of branchs in b_i^s. Then

$$\sum_{s=1}^{m_i} n_i^s \le n.$$

From Riemann-Hurwitz formula it follows that

$$2(g-1) = -2n + \sum_{i,s}(n_i^s - 1) \le -2n + (kn - \sum_{i=1}^{k} m_i) \le -2n + k(n-1)$$

and thus

$$2(g + n - 1) \le k(n-1), \quad k \ge 2\frac{n-1+g}{n-1} = 2 + 2\frac{g}{n-1}.$$

If $k = 2$ that $g = 0$, $m_1 = m_2 = 1$ and thus (P, f) is equivalent $(\bar{C}, z^n)$, where $z^n : z \mapsto z^n$.

Further we assume that $k \geq 3$.

Let $S^n \subset S_k$ be a subgroup of index n and $\psi \in T_k^*$. Put

$$\tilde{P} = \Lambda/\psi(S^n), \quad \tilde{C} = \Lambda/\psi(S_k).$$

The inclusion $S^n \subset S_k$ gives a holomorphic covering $\tilde{f} : \tilde{P} \to \tilde{C}$. Let P and $\bar{C}$ be natural compactifications of the punctured surfaces $\tilde{P}$ and $\tilde{C}$. Then $\tilde{f}$ has a continuation $f : P \to \bar{C}$ and (P, f) is a meromorphic function. Put

$$\Psi_{S^n}(\psi) = (P, f).$$

<u>Lemma 3.</u> *Let (P, f) be a meromorphic function. Then there exists a subgroup $S^n \subset S_k$ and $\psi \in T_k$ such that*

$$\Psi_{S^n}(\psi) = (P, f).$$

<u>Proof:</u> Let a_i be the singular values of (P, f). Put

$$\tilde{P} = P \backslash \cup f^{-1}(a_i), \quad \tilde{C} = \bar{C} \backslash \cup a_i \quad \text{and} \quad \tilde{f} = f \,|_{\tilde{P}} \,.$$

Then

$$\tilde{f} : \tilde{P} \to \tilde{C}$$

is a covering without singular points. From Lemma 2 it follows that there exist $\psi \in T_k$ and a covering $\varphi_\Lambda : \Lambda \to \tilde{C} = \Lambda/\psi(S_k)$. Thus there exists $\tilde{\Gamma} \subset \psi(S_k)$ such that $\varphi_\Lambda = \tilde{f}\tilde{\varphi}_\Lambda$, where $\tilde{\varphi}_\Lambda : \Lambda \to \tilde{P} = \Lambda/\tilde{\Gamma}$ is the natural projection. Put $S^n = \psi^{-1}(\tilde{\Gamma}) \subset S_k$. Then $\tilde{\Gamma} = \psi(S^n)$ and $\Psi_{S^n}(\psi) = (P, f)$.

<u>Lemma 4.</u> *Suppose that the meromorphic functions (P, f) and (P', f') have the same topological type and $(P, f) = \Psi_{S^n}(\psi)$, where $S^n \subset S_k$, $\psi \in T_k$. Then there exists $\psi' \in T_k$ such that*

$$(P', f') = \Psi_{S^n}(\psi').$$

<u>Proof:</u> From Lemma 3 it follows that there exists $S_1^n \subset S_k$ and $\psi_1' \in T_k$ such that $(P', f') = \Psi_{S_1^n}(\psi_1')$. Put

$$\tilde{P} = \Lambda/\psi(S^n), \quad \tilde{P}' = \Lambda/\psi_1'(S_1^n), \quad \tilde{C} = \Lambda/\psi(S_k), \quad \tilde{C}' = \Lambda/\psi_1'(S_k).$$

Since (P, f) and (P', f') have the same topological type, there exists a homeomorphism $\varphi_\Lambda : \Lambda \to \Lambda$ such that

$$\psi_1'(S_k) = \varphi_\Lambda \psi(S_k)\varphi_\Lambda^{-1} \quad \text{and} \quad \psi_1'(S_1^n) = \varphi_\Lambda \psi(S^n)\varphi_\Lambda^{-1}.$$

The homeomorphism φ_Λ gives rise to the isomorphism

$$\varphi_\psi : \psi(S_k) \to \psi'(S_k)$$

where $\varphi_\psi(\psi(s)) = \varphi_\Lambda \psi(s) \varphi_\Lambda^{-1}$ for $s \in S_k$. Put

$$h = (\psi_1')^{-1} \varphi_\psi \psi : S_k \to S_k \quad \text{and} \quad \psi' = \psi_1' h.$$

Thus

$$h(S^n) = S_1^n \quad \text{and} \quad (P', f') = \Psi_{S_1^n}(\psi_1') = \Psi_{h(S^n)}(\psi_1' h) = \Psi_{S^n}(\psi').$$

<u>Theorem C</u>. *Let (P, f) be a meromorphic function with k singular values. Let H be the space of all meromorphic functions of the same topological type as (P, f). Then*

$$H = T_k/Mod \cong R^{2k-6}/Mod,$$

where $Mod \subset Mod_k$. Moreover, if $(P, f) = \Psi_S(\psi)$ then

$$Mod = \{h \in Mod_k \mid h(S) = S\}.$$

<u>Proof</u>: From Lemma 3 it follows that there exists $S \subset S_k$ such that $(P, f) = \Psi_S(\psi)$. From Lemma 4 we have that $H \subset \Psi_S(T_k)$. Theorem A implies that T_k is a connected space. Moreover, from description of Ψ_S it follows that functions $\Psi_S(\psi_1)$ and $\Psi_S(\psi_2)$ have the same topological type if ψ_1 and ψ_2 are nearby. Thus

$$\Psi_S(T_k) \subset H \quad \text{and} \quad H = \Psi_S(T_k).$$

Moreover, $\Psi_S(T_k) = T_k/Mod$, where

$$Mod = \{\alpha \in Mod \mid \alpha S = \beta S \beta^{-1}, \beta \in S_k\}.$$

Thus, according to Theorem A

$$H = T_k/Mod \cong R^{2k-6}/Mod.$$

<u>Remark</u> Theorem C has an evident generalization to the situation when $f : P \to \widetilde{P}$ is a holomorphic map of finite degree, $f(P) = \widetilde{P}$ and $\widetilde{P}$ is an arbitrary surface with finitely generated group. Additional technical components of the proof follow from [8, §2].

References

1. V.I.Arnold. Topological classification of complex trigonometrical polynomials and a combinatorics of graphs with an fixed number of vertices and edges, Funct. Analysis and Applications 30 (1996), 1-14.

2. S.Cordes, G.Moore, S.Ramgoolam. Large N 2D Yang-Mills theory and topological string theory, Commun. Math.Phys. 185 (1997), 543-619.

3. B.Dubrovin. Geometry of $2D$ topological field theories, Lecture Notes in Math. 1620, Springer, Berlin (1996)

4. R.Fricke, F.Klein. Vorlesugen über die Theorie der automorphen Funktionen. V.1, Teubner, Leipzig (1897); V.2, Teubner, Leipzig (1912) (Reprinted by Johnson Reprint Corp., New York and Teubner Verlagsgesellschaft, Stuttgart (1965)

5. D.Gross, D.Taylor. Two-dimensional QCD is a string theory, Nucl. Phys., B 400 (1993), 161-180.

6. J.Harris, D.Mumford. On the Kodaira dimension of the moduli space of curves, Invent. Math. 67 (1982), 23-86.

7. A.Hurwitz. Über Riemannsche Flächen mit gegeben Verzweigunspuncten, Math.Ann, 38 (1891), 1-61.

8. S.M.Natanzon. Moduli spaces of real curves. Trans.Moscow Math.Soc. (1980) N 1, 233-272.

9. S.M.Natanzon. Spaces of real meromorphic functions on real algebraic curves, Soviet. Math. Dokl., 30:3 (1984), 724-726.

10. S.M.Natanzon. Uniformization of spaces of meromorphic functions. Soviet. Math. Dokl., 33:2 (1986), 487-490.

11. S.M.Natanzon. Topology of 2-dimensional coverings and meroporphic functions on real and complex algebraic curves, Sel. Math. Sov., 12:3 (1993), 251-291.

12. S.M.Natanzon. Spaces of meromorphic functions on Riemann surfaces, Amer. Math. Soc. Transl. (2) Vol.180 (1997), 175-180.

13. S.M.Natanzon, V.Turaev. A compactification of the Hurwitz space, Topology, 38, (1999), 889-914.

Moscow State University
Moscow, Russia.
natanzon@mccme.ru